CANADIAN BULLETIN OF FISHERIES AND AQUATIC SCIENCES 220

Chemical Oceanography in the Gulf of St. Lawrence

EDITED BY PETER M. STRAIN

Marine Chemistry Division
Physical and Chemical Sciences Branch
Department of Fisheries and Oceans
Bedford Institute of Oceanography
P.O. Box 1006, Dartmouth, N.S. B2Y 4A2

Scientific Excellence
Resource Protection & Conservation
Benefits for Canadians

DEPARTMENT OF FISHERIES AND OCEANS
OTTAWA 1988

The *Canadian Bulletins of Fisheries and Aquatic Sciences* are designed to interpret current knowledge in scientific fields pertinent to Canadian fisheries and aquatic environments.

The *Canadian Journal of Fisheries and Aquatic Sciences* is published in annual volumes of monthly issues. *Canadian Special Publications of Fisheries and Aquatic Sciences* are issued periodically. These series are available from authorized bookstore agents and other bookstores, or you may send your prepaid order to the Canadian Government Publishing Centre, Supply and Services Canada, Ottawa, Ont. K1A 0S9. Make cheques or money orders payable in Canadian funds to the Receiver General for Canada.

Communications Directorate

Nicole Deschênes	Director General
John Camp	Director
Gerald J. Neville	Editorial and Publishing Services

Typesetter: Typoform
Printer: Kromar Printing Ltd., Winnipeg
Cover Design: Miracom Communications Inc., Ottawa

Cat. No. Fs 94-220E   ISBN 0-660-13078-5
DFO / 4160   ISSN 0706-6503

Correct citation for this publication:

STRAIN, P.M. [ED.]. 1988. Chemical oceanography in the
     Gulf of St. Lawrence. Can. Bull. Fish.
     Aquat. Sci. 220: 190 p.

Contents

Abstract

STRAIN, P. M. [ED.] 1988. Chemical oceanography in the Gulf of St. Lawrence. Can. Bull. Fish. Aquat. Sci. 220: 190 p.

This book is a review, synthesis, and addition to data on the chemical oceanography and geochemistry of the Gulf of St. Lawrence, the St. Lawrence Estuary, and the Saguenay Fjord. Since it has been produced by scientists in the Marine Chemistry Division of Physical and Chemical Sciences Branch, Department of Fisheries and Oceans, Canada, its focus is on research conducted out of the Bedford Institute of Oceanography but it also includes discussion of work conducted by other agencies on the Gulf of St. Lawrence.

The first part of the book is an examination of the basic chemical oceanography of the Gulf. The geography, physical oceanography, and geology of the region are briefly described. Existing data on suspended particulate matter (SPM) is reviewed, new data is presented and a budget for SPM in the Gulf is developed. New nutrient data are presented for the Upper St. Lawrence Estuary and the Gulf source waters in the open Atlantic east of Cabot Strait. Nutrient cycling and regeneration processes and nutrient budgets are discussed. New data on stable oxygen isotope ratios of waters in the Gulf are discussed and the potential for using such measurements to identify fresh water sources is explored.

The book brings together existing data on dissolved and particulate organic carbon in the water column and the organic fraction of surficial sediments. The distributions of these organic fractions are described and a budget for the Gulf is constructed. The carbon isotope ratio work that has been used to track terrestrial and riverine organic matter in surficial sediments and to study interactions between inorganic carbon, particulate organic carbon and planktonic carbon is reviewed. The results of a 5-year program monitoring organic carbon inputs from the St. Lawrence River are described.

The behaviour of trace metals during estuarine mixing, the interactions between dissolved and particulate phases, and the influence of diagenetic processes in sediments on metal distributions in the water column are described. New data for the Saguenay Fjord are analyzed. Models describing the fluxes of metals through the coastal zone are discussed. Research on the behaviour and distributions of trace metals in surficial sediments in the Gulf is reviewed. The levels of metals in Gulf sediments are compared to those in other coastal areas and the relationships between trace metals and sediment composition are examined to reveal details on the transport and deposition modes of different trace metal fractions.

Also, the book is a description of research on pollution in the Gulf of St. Lawrence. The use of accurately dated cores from the Saguenay Fjord for the study of both pollution inputs and the history of climate variations is reviewed. The history of mercury pollution in the Saguenay Fjord is examined and the deposition history of radionuclides is used to model metal transport processes in the Saguenay drainage basin. The limited information available on the incidence of organic contaminants in the Gulf is reviewed. The results of a 10-year survey of petroleum hydrocarbons in the waters of the Gulf are reported. Oil residue concentrations declined during this period due to decreasing fluxes from the open Atlantic to the Gulf.

The final chapter is an examination of the different approaches that have been used in data collection and interpretation in the study of chemical processes. The scope of the present understanding of marine chemistry is evaluated, limitations in existing data identified, and recommendations are made for new approaches in future research. An appendix cataloging chemical data on the Gulf of St. Lawrence available at BIO concludes the work.

Résumé

STRAIN, P. M. [ÉD.] 1988. Chemical oceanography in the Gulf of St. Lawrence. Can. Bull. Fish. Aquat. Sci. 220: 190 p.

Cet ouvrage revoit, résume et présente les données nouvelles sur l'océanographie chimique des eaux du golfe du Saint-Laurent, de l'estuaire du Saint-Laurent et du Saguenay. Rédigé par des scientifiques de la Division de la chimie marine de la Direction générale des Sciences physiques et chimiques du ministère des Pêches et des Océans du Canada, il porte surtout sur les recherches effectuées à l'Institut océanographique de Bedford, mais aussi sur les travaux entrepris par d'autres organismes dans le golfe du Saint-Laurent.

La première partie traite de l'océanographie chimique fondamentale du golfe. La géographie, l'océanographie physique et la géologie de la région font l'objet d'une brève description. Les données existantes sur les matières solides en suspension (MSS) sont examinées, de nouvelles données sont intégrées et un bilan des MSS est établi pour le golfe. De nouvelles données sur les matières nutritives dans le bassin supérieur de l'estuaire du Saint-Laurent, dans les eaux de l'Atlantique qui alimentent le golfe à l'est du détroit de Cabot, sont présentées. Le cycle, le processus de régénération et le bilan des matières nutritives sont étudiés. De nouvelles données sur les rapports des isotopes stables de l'oxygène dans les eaux du golfe sont examinées et l'utilisation potentielle de ces mesures pour identifier les sources d'eau douce est examinée.

L'ouvrage rassemble les données existantes sur le carbone organique dissous et particulaire dans la colonne d'eau, ainsi que sur la fraction organique des sédiments de surface. La distribution de ces fractions est décrite et un bilan est établi pour les eaux du golfe. Le rapport isotopique du carbone utilisé pour déceler la matière organique d'origine terrestre et fluviale dans les sédiments de surface et étudier les interactions entre le carbone inorganique, organique particulaire et planctonique est examiné. Les résultats d'un programme quinquennal de surveillance des sources de carbone organique dans le fleuve Saint-Laurent sont décrits.

Le comportement des métaux-traces pendant le mélange des eaux estuariennes, les interactions entre les phases de dissolution et de formation des particules, et l'influence des diagenèses dans les sédiments sur la distribution des métaux dans la colonne d'eau sont décrits. De nouvelles données recueillies sur le Saguenay sont examinées et des modèles du transport des métaux dans la zone côtière sont étudiés. Les recherches sur le comportement et la distribution des métaux-traces dans les sédiments de surface du golfe font l'objet d'une révision. Les concentrations de métaux dans les sédiments du golfe et d'autres zones côtières sont comparées, et les relations entre les métaux-traces et la composition des sédiments sont examinées afin de mieux connaître les processus de transport et de sédimentation des différentes fractions des métaux-traces.

L'ouvrage décrit également les recherches effectuées sur la pollution dans le golfe du Saint-Laurent. L'utilisation de carottes, prélevées dans le fjord du Saguenay et datées avec précision, dans l'étude des sources polluantes et des variations climatiques est examinée. La pollution par le mercure dans le fjord du Saguenay est étudiée, et le dépôt des radionucléides est utilisé pour modéliser les processus de transport des métaux dans le bassin hydrographique du Saguenay. Les données sur les contaminants organiques présents dans les eaux du golfe sont examinées. Les résultats d'une étude décennale sur les hydrocarbures du pétrole dans les eaux du golfe sont présentés. Les concentrations d'hydrocarbures résiduels ont diminué dans le golfe durant cette période en raison de la réduction des apports en provenance de l'Atlantique.

Le dernier chapitre porte sur les différentes méthodes de collecte et d'interprétation des données dans l'étude des processus chimiques. L'état actuel des connaissances de la chimie marine est évalué, les limitations des données existantes sont définies et des recommandations sont formulées quant aux nouvelles orientations des recherches futures. Enfin, les données recueillies par l'IOB sur la chimie du golfe du Saint-Laurent sont présentées en annexe.

Foreword

In the early 1970's coastal and estuarine studies were, to some extent, unfashionable. The major investment of marine chemical effort globally was on the deep oceans, as exemplified by the Geochemical Ocean Sections (GEOSECS) Program. Towards the late 70's, scientific interest in coastal areas substantially increased partly as a result of the growing awareness of inshore environmental problems and the recognition that there were still many unanswered questions about processes in estuaries and coastal marine environments.

Most of the research that is discussed in this book was conducted out of the Bedford Institute of Oceanography. It began in 1970 when the Chemical Oceanography Division of the Department of Energy, Mines and Resources (now the Marine Chemistry Division of the Department of Fisheries and Oceans) was created at BIO. There was a need at that time to find a suitable focus for the developing work in chemical oceanography and the Gulf of St. Lawrence was identified as a convenient and significant area for the study of estuarine and coastal processes. The economic importance of the Gulf to Canada gave additional impetus to its choice as a focus for a variety of marine scientific studies. Gulf studies continued to dominate the work of the Division until about 1980 by which time interests had broadened considerably to include work in the Canadian Arctic and elsewhere in the North Atlantic. The opening of the Institut Maurice Lamontagne at Ste. Flavie, Quebec, in 1987 provided a new facility for coastal and estuarine research in eastern Canada. Most new research in the Gulf will be conducted at this Institut, which should be fully operational in 1991.

This book is a convenient summary of work on the Gulf carried out by the Chemical Oceanography Division prior to the transfer of responsibilities to the Institut Maurice Lamontagne. It partly reflects the "coming of age" of chemical oceanography as a distinct oceanographic discipline, both in Canada and abroad, during the last two decades. In addition, it exemplifies the broadening of marine chemical science from the study of only those processes occurring within the ocean to include the study of the processes and rates of material injection from terrigenous and atmospheric sources. The linkages between physical, chemical, and biological marine sciences are becoming more important as more is learnt about different marine processes and the interactions between them. The fact that this book primarily contains information on chemical oceanographic investigations is far more a reflection of history than of a conviction that the disciplines can proceed independently. The challenge for the future is to increase the level of multi-disciplinary scientific effort in the study of regions like the Gulf of St. Lawrence.

J.M. BEWERS
Head, Marine Chemistry Division
Physical and Chemical Sciences Branch
Department of Fisheries and Oceans

Acknowledgements

The chemical oceanographic programs conducted out of the Bedford Institute of Oceanography that form the foundation for this book were conducted over the period from 1970 to the present. A large number of people have contributed to these programs. Alan Walton, Head of the Chemical Oceanography Division from 1970 to 1978, established many of the programs whose results are discussed in this book. J.M. Bewers, Head of the Division since 1978, directed the later period of Gulf research, and was an active participant. The late A.R. Coote, C.D.W. Conrad, and W. Young were all active in Gulf programs until their deaths. Art Coote was one of the first to be curious about the dynamics of the Gulf as a system.

In addition to those mentioned above and the authors of the chapters, contributors include:

J.H. Abriel	K.M. Ellis	L. Morash
B.P. Amirault	W.D. Fraser	R.W.P. Nelson
J. Bates	J. Guilderson	D.W. Pottie
J.L. Barron	G.T. Hagell	R.T.T. Rantala
F.J. Bishop	W.R. Hardstaff	S. Roy
Cai Deiling	S. Hartling	J.B. Simms
J.A. Campbell	M. Hartwell	R.D. Smillie
N.F. Crewe	R.S. Hiltz	Song Shulin
C.C. Cunningham	M. Huh	B. Sundby
J.A. Dalziel	J.D. Leonard	M. Tucker
R. Decoste	I.D. Macauley	R.W. Walker
R. Dobson	C.D. MacGregor	L.R. Webber

The success and breadth of these research programs has also been dependent on many additional summer students and other temporary employees.

A number of people deserve thanks for their contributions to the preparation of this monograph. C. E. Cosper co-ordinated the project in its early stages. P. A. MacPherson drew most of the figures. Twenty busy researchers took the time to review chapter manuscripts.

CHAPTER I

Chemical Oceanography in the Gulf of St. Lawrence: Geographic, Physical Oceanographic, and Geologic Setting

P. M. Strain

Marine Chemistry Division, Physical and Chemical Sciences Branch,
Department of Fisheries and Oceans, Bedford Institute of Oceanography,
P.O. Box 1006, Dartmouth, N.S. B2Y 4A2

Introduction

The Gulf of St. Lawrence (Fig. I.1) is a major marginal sea on the eastern seaboard of Canada. It is bounded by the province of Quebec and three of the four Atlantic provinces — New Brunswick, Nova Scotia, and Newfoundland. The fourth Atlantic province — Prince Edward Island — is located in the Gulf. The Gulf has figured significantly in Canadian history. Vikings ventured at least to the edge of the Gulf ca. 1000 A.D. John Cabot apparently explored both Cabot Strait and the Strait of Belle Isle in 1497, although the exact track of his voyage is in doubt. The first European known to have entered the Gulf was Jacques Cartier who, in 1534, explored as far west as Anticosti Island and, in the next year, sailed to Quebec and up the St. Lawrence River to Montreal. The Gulf continued to serve as the major route to central Canada for the next 300 years and remains an important transportation avenue to the industrial centre of North America today. Its natural resources include the fishery, whose landings account for approximately 25 % by weight of the total Canadian fishery (Dep. of Fisheries and Oceans 1987). The southern shores of the Gulf are active recreational sites in summer, and support an important tourism industry. Thus the Gulf has a profound impact on the economic and social life of eastern Canada.

This book will describe chemical oceanographic research that has been conducted on the Gulf of St. Lawrence system, concentrating on the work performed by scientists

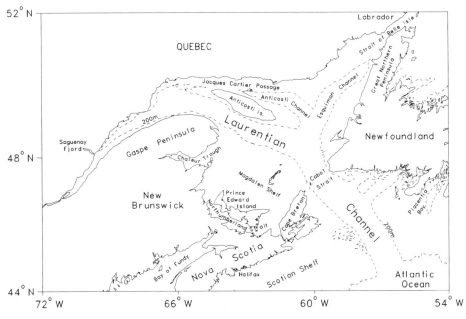

FIG. I.1. The Gulf of St. Lawrence. The dashed line is the 200 m depth contour.

at the Bedford Institute of Oceanography during the 1970's and early 1980's. The rest of this chapter will describe aspects of the Gulf's physical setting, its physical oceanography, and its geologic setting that are relevant to an understanding of the chemical oceanography of the region.

Physical Setting

In this publication, the term "Gulf of St. Lawrence" is used to describe the entire estuarine system from the mouths of the St. Lawrence and Saguenay rivers in the west to the two connections with the open Atlantic Ocean in the east — Cabot Strait and the Strait of Belle Isle (Fig. I.1-2). The Gulf of St. Lawrence has an area of 226×10^3 km^2, making it approximately half as large as the Baltic or Black Seas. With an average depth of 152 m, it contains a volume of 34.5×10^3 km^3, one third more than the Baltic Sea.

For its area, the Gulf receives runoff from a larger drainage basin (1.51×10^6 km^2) than any other semi-enclosed sea. Much of this freshwater input arrives via the St. Lawrence River, which, in terms of water flow, is the second largest river in North America (after the Mississippi). The St. Lawrence discharges ≈ 350 km$^3 \cdot$yr^{-1} at Quebec City (this is an average for the period 1960-72, Pocklington 1982). Quebec City also marks the upstream limit of the St. Lawrence Estuary (Fig. I.3), which is here defined as the region between Quebec City and Pointe des Monts (a distance of ≈ 400 km). (The phrase "open Gulf" will frequently be used to describe the Gulf exclusive of the Estuary.) The Estuary is further subdivided into the Upper Estuary (≈ 180 km long, and reaching a width of 24 km at its downstream end), between Quebec and the mouth of the Saguenay Fjord, and the Lower Estuary, between the Saguenay and Pointe des Monts. (This is the most common, but not the only, scheme for subdividing the Estuary — d'Anglejan (1981) refers to Upper, Middle, and Lower Estuaries which

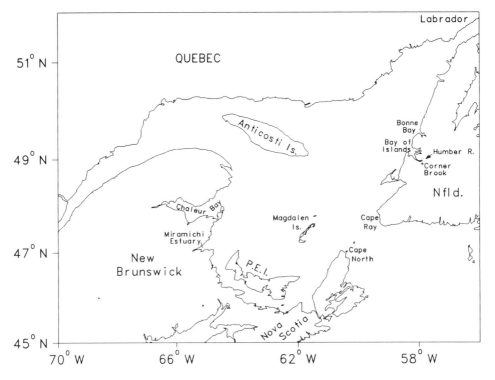

FIG. I.2. The Gulf of St. Lawrence, showing some additional places referred to in the text.

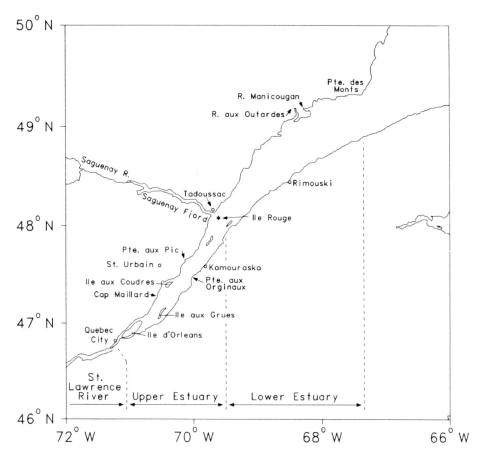

FIG. I.3. The St. Lawrence Estuary and the Saguenay Fjord.

correspond to the St. Lawrence River, the Upper Estuary, and the Lower Estuary as used here; Greisman and Ingram (1977) refer to the Lower Estuary as the Maritime Estuary.)

Other major sources of freshwater inputs to the Gulf of St. Lawrence are located to the north and west of the Gulf. The Saguenay River, which is connected to the Estuary through the 95 km long Saguenay Fjord, contributes an additional 40 km^3·yr^{-1} and drains an area of 78,000 km^2; the combined flow of the Manicouagan and Aux Outardes rivers, which flow into the Lower St. Lawrence Estuary west of Pointe des Monts, is 32 km^3·yr^{-1} (averages for 1960-70, from data in Canadian National Committee 1972). A large number of major rivers drain into the open Gulf from the north shore, with a combined average flow of 130 km^3·yr^{-1} (1960-70, Jordan 1973). In contrast, freshwater inputs around the rest of the Gulf — from Newfoundland, the Gaspé, New Brunswick, Nova Scotia, and Prince Edward Island — total only 47 km^3·yr^{-1} (1965-70, Jordan 1973).

Based on the above data, the contribution of the St. Lawrence River (measured at Quebec City) to the total freshwater flux into the Gulf is a maximum of 58 % (a maximum because some smaller rivers emptying into the St. Lawrence Estuary have not been included). Since different averaging periods have been used for the different inputs, this estimate is an approximation only. In addition, caution must be used in comparing this calculation with previous reports — in at least some cases the St. Lawrence input has apparently been defined as the total freshwater entering the open Gulf from the St. Lawrence Estuary. For example, Dickie and Trites (1983), quoting El-Sabh (1977), report that the St. Lawrence supplies 75 % of the Gulf's freshwater. El-Sabh's (1977) actual fraction was two-thirds — but the 75 % estimate is close to the 70 % calculated from the data presented here for the combined flows of the St. Lawrence, the Saguenay, the Aux Outardes, and the Manicouagan rivers.

The most conspicuous feature of the bathymetry of the Gulf is the Laurentian Channel. This is a deep, steep-sided trough which extends more than 1 200 km from just east of the mouth of the Saguenay Fjord in the St. Lawrence Estuary through the central Gulf and Cabot Strait to the edge of the continental shelf outside the Gulf of St. Lawrence. At its upstream limit, maximum depths increase from less than 40 m to greater than 320 m over a distance of less than 20 km. The maximum depth increases steadily to 535 m just west of Cabot Strait, although it shoals to 480 m in the Strait itself. The Laurentian Channel has an average width of ≈ 50 km, with a maximum width of ≈ 80 km. Two more troughs branch from the Laurentian Channel in the northeastern part of the Gulf — the Esquiman Channel that parallels the west coast of Newfoundland, and the Anticosti Channel, in turn connected to the Esquiman Channel, which parallels the Laurentian Channel north of Anticosti Island (see Fig. I.1). If the 200 m contour line is used to define the extent of the deep channels, then about half of the Gulf's area overlies deep water.

The Gulf of St. Lawrence is a drowned Paleozoic lowland whose fluvial drainage system was modified during the Pleistocene glaciations. Glacial erosion widened, deepened and straightened the fluvial valleys, producing the Laurentian Channel system with steep sides and broad, hummocky floors.

North of the Laurentian Channel, and in the area between the Laurentian Channel and the Gaspé peninsula (including the Lower St. Lawrence Estuary), the bottom shoals rapidly from the deep channels to the mainland or Anticosti Island. The southeastern part of the Gulf, however, is a broad, shallow (usually < 75 m) shelf known as the Magdalen Shelf. A similar, but narrower, shelf is found between the Esquiman Channel and the west coast of Newfoundland. Approximately one quarter of the Gulf overlies water depths of < 75 m. Three major islands are found in the Gulf — Anticosti Island, Prince Edward Island, and the Magdalen Island chain.

The Upper St. Lawrence Estuary is relatively shallow, with a maximum depth of ≈ 30 m in the upper two-thirds of its length, with greater depths found on the north side of the Estuary. However, even there dredging is necessary to maintain a ship channel with a minimum depth of 12.5 m. The southern part is shallower, composed of a complex pattern of channels and flats which form low islands and emergent banks. The lower third of the Upper Estuary, known as the eastern basin, reaches depths of 120 m.

The Saguenay Fjord consists of deep basins separated from each other and from the St. Lawrence Estuary by shallow sills. The inner (western) basin, which extends some two thirds of the length of the Fjord, has depths reaching 275 m. The important sills include an 80 m sill located ≈ 18 km above the mouth of the Fjord and the main sill (maximum depth = 20 m) found at the mouth. The outer basin, between these two sills, reaches depths of 250 m. Several much smaller fjords, including the Bay of Islands and Bonne Bay, are found on the west coast of Newfoundland.

The principal connection between the Gulf and the Atlantic is Cabot Strait, which has a maximum depth of 480 m, is 104 km wide, and has a cross-sectional area of 35 km². A second connection is the Strait of Belle Isle, which narrows to 16 km and has a sill depth of only 60 m.

Climate on land areas adjacent to the Gulf of St. Lawrence varies from temperate (humid continental, cool summer, no dry season) along the southern boundaries of the Gulf to subarctic along the northern shore of the open Gulf. The entire area has long, snowy winters (150-375 cm snow, 80-200 days of snow cover) with average January temperatures between − 15 and − 4°C. Summers are cool to mild having 1-4 months with mean temperatures > 10°C. Annual precipitation amounts range from 75 to 140 cm.

This annual climate pattern has considerable implications for freshwater inputs to the Gulf and for the stratification and surface circulation of Gulf waters (see below). In addition, it results in the complete freezing of Gulf surface waters. Ice formation begins in late December to early January in the St. Lawrence Estuary, on the western and southern edges of the Magdalen Shelf, and in the Strait of Belle Isle (Matheson 1967; Forrester and Vandall 1968). By mid-January the Estuary and the western half and northern shore of the Gulf contain high concentrations of ice. By the end of January, ice is present throughout the Gulf. Consolidation of the pack continues to the end of February, by which time the entire Gulf has 7-9 tenths cover. From this point, ice density decreases, but appreciable open water does not occur until the early part of April, by which time the Estuary east to Anticosti Island is clear, as is an area extending into the Gulf from the northern part of Cabot Strait. The central part of the Gulf is completely open by the end of April, but remnants of the pack can be found near Prince Edward Island and in the northeast corner of the Gulf until the end of May. Thus, the freshwater held in Gulf ice is released during a period of 2-3 months in the spring.

There are limited, and widely varying, estimates of the amount of freshwater tied up in Gulf of St. Lawrence ice each winter. Forrester and Vandall (1968) used weekly ice summary charts for the Gulf of St. Lawrence (Ice Forecasting Central, Canadian Department of Transport) to inventory ice in the Gulf. These charts do not give explicit information on ice thicknesses, but instead list the fractions of ice in "new", "young", "medium winter", and "thick winter" categories. Forrester and Vandall used "scattered reports of ice thicknesses from icebreakers" to assign the ranges < 5, 5-15, 15-30, and > 30 cm, respectively, to thicknesses in these classes. For the 6-year period 1962-67, based on these assumptions, they calculated that the mean volume of ice peaked at 35.6 km^3 at the end of February. They also calculated that maximum average ice thicknesses never exceeded 25 cm in any of their 10 subdivisions of the Gulf; in most areas it was between 15 and 20 cm. Even after correcting for the 7-9 tenths cover at the peak of the ice season, it is difficult to rationalize these thicknesses with the four ice cores discussed in Chapter V, all of which were longer than 40 cm and one of which was longer than 80 cm.

Fairbanks (1982) also made an estimate of the volume of freshwater contained in Gulf ice. Using very low resolution ice charts from the U.S. Navy (they cover the area from 40-90 °N and 15 °E to 105 °W), he estimated that 280 km^3 of water is tied up in Gulf ice. This amount of ice would correspond to an average ice thickness of 1.2 m over the entire area of the Gulf and is equivalent to 80 % of the annual discharge of the St. Lawrence River at Quebec.

Man's activities have significantly altered some aspects of the Gulf of St. Lawrence environment. The most noticeable effect is the regulation of freshwater discharges by the construction of the St. Lawrence seaway and of major hydroelectric developments on many of the rivers in Quebec. From a chemical standpoint, the Gulf receives waters that have been subjected to large industrial and domestic waste discharges during their passage through the heavily industrialized areas surrounding the Great Lakes and the St. Lawrence River valley. Industrial wastes also derive from the concentration of heavy industry — pulp mills, aluminum refineries, and mining smelters — found along the Saguenay River (see Fig. VIII. 1). Locally important industrial discharges also occur at sites around the open Gulf, such as the large pulp and paper mill discharging into the Bay of Islands and pulp and mining operations in northern New Brunswick.

Physical Oceanography

Figure I.4 shows a typical salinity distribution for the St. Lawrence Estuary. The upstream limit of salt intrusion is near Île d'Orléans. The Upper Estuary varies from being vertically homogeneous, i.e. completely mixed, at its upstream end, to being well stratified over the deeper waters in the eastern basin. Through most of its length,

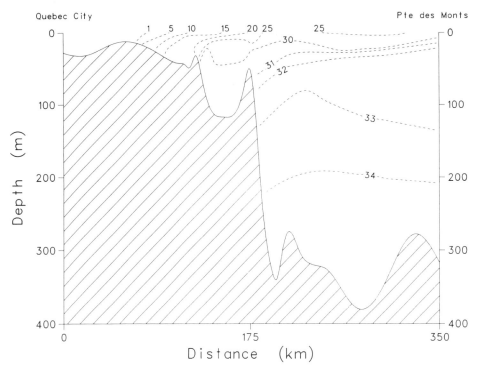

FIG. I.4. A typical longitudinal cross-section of the salinity distribution in the St. Lawrence Estuary (redrawn from Bewers and Yeats 1979, with permission).

the Upper Estuary is partially mixed, as expected for an estuary that has both a large freshwater input and strong tidal mixing. The Lower Estuary is well stratified through-out its length.

At first glance, the flow in the Upper Estuary follows the classical two-layer estuarine description, in which the entrainment of saline water into the outflowing freshwater surface layer is balanced by a net residual landward flow in the deep saline layer. Such a model does describe some of the major features of the circulation in the Gulf of St. Lawrence, such as the flow of deep water through Cabot Strait along the Lauren-tian Channel to the mouth of the Saguenay. In addition, it explains the presence of a feature that dominates the chemistry of the Upper Estuary — a turbidity maximum zone. This zone contains suspended matter at higher concentrations than in either the river or in the saltwater present in the eastern basin and further downstream. Postma (1967) explained the maintenance of turbidity maxima by describing the ability of such a flow regime to trap suspended matter which settles out of the outflowing water into the inflowing water. In reality this picture is made much more complex by the interac-tion of strong tides with the complex bathymetry of the Upper Estuary, which leads to a marked reduction of stratification within the turbidity maximum at times of low freshwater input (d'Anglejan 1981). In addition, Kranck (1979) has stressed the impor-tance of flocculation processes in trapping particles in the turbidity maximum that would otherwise settle too slowly to sink out of the surface outflowing layer. Flocculation occurs more readily with organic matter, which is therefore preferentially trapped in the tur-bidity maximum, than with inorganic material. (See Chapter II for further discussion on the turbidity maximum).

Circulation in the Estuary is driven by freshwater inputs, wind and the tide. Fresh-water inputs show marked variations, both seasonally and interannually. Some recent data (1981–1984, Pocklington and Tan 1987) show that St. Lawrence River discharges are $\approx 18 \times 10^3$ m$^3 \cdot$s^{-1} at their peak in April–May, compared to

$\approx 11 \times 10^3$ m^3·s^{-1} at their lowest in January–February. The seasonal variability is lower than in many other rivers because of damping due to the large storage capacity of the St. Lawrence drainage basin. The spring freshet has been reduced in recent decades by large hydroelectric power projects and other water level control facilities.

Mean tides in the Upper Estuary reach amplitudes of 5 m in some areas, with large tides up to 7 m. The freshwater discharge at Quebec produces markedly assymetric tides with long ebbs and fast floods in the upper reaches of the Upper Estuary (d'Anglejan 1981). Tidal excursions in the relatively deep water on the north side of the Estuary are greater than, and ahead of, those in the shallow water on the south side. Considerable phase lags between tidal heights and currents (low water occurs at the late ebb stages) are indicative of the extensive dissipation of tidal energy by bottom friction. Sediment resuspension, an important feature of Upper Estuary sediment dynamics, is driven by this energy. Fine sediments accumulate only in isolated pockets.

The Saguenay Fjord is a fjord class estuary (Dyer 1973) in which the freshwater discharge is confined to a shallow surface layer (10–15 m) separated from deeper waters by an intense halocline. The salinity of the surface layer increases from 0 at the head to ≈ 18 at the mouth of the Fjord. The halocline prevents direct chemical interaction between Saguenay sediments and the surface layer, except at the head of the Fjord.

Deep water in the Fjord is relatively homogeneous, with salinities between 29 and 30.5 throughout the year, and temperatures between 0 and 5 °C, with some seasonal variability. Therriault and Lacroix (1975) suggested that this deep water is derived from surface water (i.e. above sill depth) in the St. Lawrence Estuary, but Seibert et al. (1979) pointed out that the temperature and salinity characteristics of the water are inconsistent with such a hypothesis. They suggested that internal tides are responsible for lifting water with the appropriate T–S properties from its "normal" depth of 60–70 m up to sill depth, which then flows downslope mixing with Fjord water by entrainment as it enters the outer basin. Internal waves with amplitudes up to 80 m have been observed in this area (El-Sabh 1979). They are likely generated by the interaction between the barotropic tide and the rapid shoaling at the head of the Laurentian Channel in the Lower Estuary. Both the inner and outer basins of the Saguenay show the tidal frequency variations consistent with such a mechanism.

The resulting replacement time for deep waters in the Fjord is much shorter than usually found in sill Fjords of comparable bathymetry. Seibert et al. (1979) estimated that this time is on the order of days to weeks. The energy supplied by the tidal density flow over the sill is probably not sufficient to account for such rapid flushing — these authors suggest that internal waves within the Fjord generated by scattering of the tide at abrupt changes in topography could provide the necessary mixing energy.

From a chemical perspective, the most important consequence of the rapid replacement of deep waters in the Fjord is the relatively high levels of dissolved oxygen found in these waters. Seibert et al. (1979) reported that rarely are values less than 180 micromolar (μM) observed; indeed concentrations are commonly > 270 μM at 250 m in the inner basin.

The same tidal forces developed at the head of the Laurentian Channel which contribute to the rapid replacement of deep Saguenay waters also play a primary role in mixing deep Laurentian Channel water up into the surface layer of the Estuary. Dickie and Trites (1983) note that mixing can be so intense that the water column may be homogeneous (at 5°C and a salinity of 28) in this area even in July. Such mixing, which leads to very high surface nutrient levels in the Lower Estuary, is consistent with Steven's (1974) concept of a "nutrient pump" in this area.

The interaction of the regulation of the freshwater discharge and the mixing of deep waters with surface waters is complex and has been a subject of some controversy. Reid (1977) suggested that the internal tide mechanism may be diminished at times of high river discharge because of increased stratification at some critical sills in the Estuary near the mouth of the Saguenay. On the other hand, Neu (1982a, b) argued that the amount of deep water entrained into the surface layer of the two-layer estuarine flow was directly correlated with freshwater discharge, and that the natural coincidence of the spring freshet and other favourable growing conditions was critical to the high productivity found in many coastal regions. Bugden et al. (1982) discussed the effects of fresh water runoff throughout the Gulf. The counteracting influences of the increased shear and the increased stability that accompany increased discharge make it difficult to assess the effect of freshwater regulation in some areas. For example, they concluded that data was insufficient to assess the impact in the Estuary or at the head of the Laurentian Channel. However, they did state that increased discharge would likely increase mixing in the Gaspé current and decrease it over the Magdalen Shelf. Sinclair et al. (1986) have summarized these conclusions and extended the discussion to compare the impact of local freshwater inputs and non-local ocean circulation.

The surface low salinity discharge from the Saguenay Fjord sets to the southern side of the Estuary where it joins the low salinity water from the Upper Estuary. The Saguenay plume is displaced upstream into the Upper Estuary by flood tides. (Greisman and Ingram (1977) note that these tidal excursions are of the order of 10 km.) The surface circulation in the Lower Estuary is complex and is influenced by winds and atmospheric pressure gradients in addition to barotropic and internal tides and Coriolis effects (El-Sabh 1979). El-Sabh has proposed that the summer circulation pattern in the Lower Estuary is composed of two counter-rotating gyres to the north and a seaward flow on the south shore (Fig. I.5). By comparing the peak discharge from the St. Lawrence River and the mean salinity at Pointe des Monts, El-Sabh (1979) estimated a transit time of 1 month for freshwater in the St. Lawrence Estuary. At the mouth of the Estuary, the seaward flow on the south shore (the Gaspé current) is augmented by a cyclonic gyre found between Anticosti Island and Pointe des Monts to form an intense, narrow flow in the upper 25–50 m along the Gaspé peninsula.

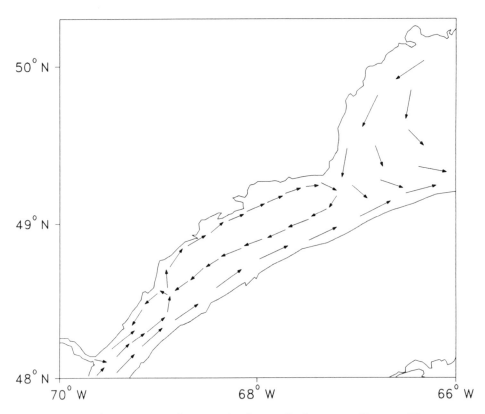

FIG. I.5. Typical summer circulation in the Lower St. Lawrence Estuary. The western part of the Anticosti gyre is at the right (redrawn from El-Sabh 1979, with permission).

The vertical structure of waters in the open Gulf changes seasonally. In winter, two distinct water layers are present: a surface mixed layer, 50 - 100 m thick, with temperatures at or just above the freezing point (− 1.5°C) and salinities between 31 and 32, and deep, warm (4-6°C) water with salinities close to 34.6 (Fig. I.6; Trites 1972). In summer, surface warming separates the upper layer into two distinct layers: a warm (15°C), thin (10-30 m), relatively fresh (S = 27-32) surface layer and an intermediate cold layer, slightly warmer (− 0.5 to 1°C) than the surface winter layer, with salinities from 31.5 to 33 (Dickie and Trites 1983). This intermediate cold layer was at one time believed to originate on the Labrador Shelf, whose waters flow into the Gulf through the Strait of Belle Isle and have very similar T-S properties. Dickie and Trites (1983) believed that the exchange through the Strait of Belle Isle is small compared to the total volume of the intermediate layer in the Gulf and that most of this layer is a remnant from the previous winter's cooling. Recent work by Petrie et al. (1988), however, shows that the Gulf receives substantial net inputs through the Strait of Belle Isle and that both *in situ* cooling and advection of Labrador Sea water contribute to the Gulf's intermediate layer.

Figure I.7 is a stylized representation of the typical summer circulation in the open Gulf. Notable features include the generally cyclonic flow within the Gulf, the bidirectional flows observed both in Cabot Strait and the Strait of Belle Isle, and the Anticosti gyre, found just to the west of Anticosti Island. The most important driving mechanisms for this circulation are the wind and the large freshwater discharge from the St. Lawrence Estuary. This circulation pattern divides the Gulf — the southern region over the Magdalen Shelf is more directly influenced by the freshwater outflow from the Estuary than is the northeastern region. By tracking the minimum seasonal salinity from the St. Lawrence River to Cabot Strait, El-Sabh (1977) estimated that the mean flushing time of fresh water in the Gulf is 3 months.

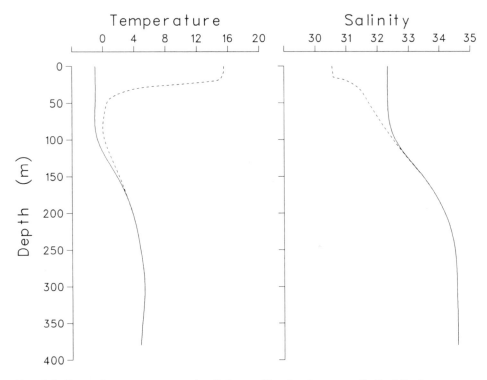

FIG. I.6. Typical temperature and salinity profiles for the open Gulf of St. Lawrence. Solid lines = winter; dashed lines = summer (redrawn from Trites 1972, with permission).

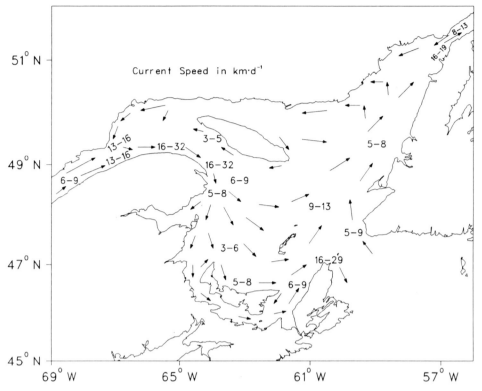

FIG. I.7. Typical summer circulation in the open Gulf of St. Lawrence. Numbers indicate surface flow in km • d^{-1} (redrawn from Trites 1972, with permission).

The deep water in the Gulf originates outside the Gulf. This water, which is contiguous with the deeper water in the Laurentian Channel outside Cabot Strait, is composed partially of water from the inshore branch of the Labrador Current. Dickie and Trites (1983) note that the maximum temperature of this layer, which varies between 4 and 6°C, is correlated with the strength of the Labrador Current. El-Sabh (1977) used these long-term temperature variations to calculate a transit time of between 2 and 2.5 years for deep water flowing from Cabot Strait to the St. Lawrence Estuary.

A number of physical oceanographic studies on water fluxes through Cabot Strait have been conducted. Trites (1972) showed cross-sections of residual currents at several locations in the Gulf, including Cabot Strait, and stated that the total seaward flux through Cabot Strait is ≈ 50 times the mean freshwater discharge. Dickie and Trites (1983) later revised this estimate to a factor of ≈ 30 based on a detailed study of flows through Cabot Strait by El-Sabh (1975, 1977). El-Sabh used geostrophic calculations, based on the available CTD data backed up by a limited number of direct current measurements, to describe in detail the strength and location of the flows in each direction through Cabot Strait (as well as at a cross-section at the mouth of the St. Lawrence Estuary). On average, ≈ 20 % of the volume outflow through Cabot Strait is located in the upper 25 m on the Cape Breton side of the Strait. Inflow from the Atlantic occurs at all depths on the Newfoundland side, and in some deeper waters in the centre of the Strait (Fig. I.8). El-Sabh also compiled detailed tables showing outflow, inflow, and net flow in each depth layer at a number of times of year, a format that is particularly useful for the calculation of chemical mass fluxes.

Several of the chapters in this book employ box models to construct budgets for materials in the Gulf (Chapters II, III, and IV for suspended particulate matter, nutrients, and organic carbon, respectively). The fluxes through Cabot Strait are major components of such models. Initial comparisons between these models revealed discrepancies (such as unrealistic concentrations of organic carbon in suspended matter) that were

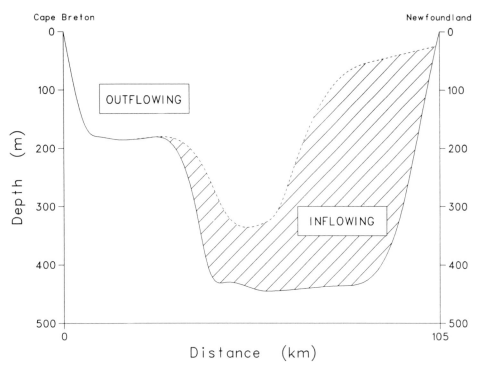

FIG. I.8. Locations of the inflowing and outflowing waters in Cabot Strait (redrawn from Dickie and Trites 1983, with permission).

traced to the use of different estimates for water fluxes through Cabot Strait. The models presented here are all based on El-Sabh's detailed description of flow through Cabot Strait, adjusted as suggested by Bugden (1981) to account for the scarcity of measurements very close to Cape Breton Island. Updating aspects of the models based on earlier studies of Cabot Strait has removed contradictions between them, but some quantities in the models may differ from previously published work.

El-Sabh's calculations assumed that exchanges through the Strait of Belle Isle are insignificant, and that the total amount of salt in the Gulf is constant. As noted above, Petrie et al. (1988) have shown that net water fluxes through the Strait of Belle Isle are much larger than some previous work had indicated. Furthermore, they note that long-term variations in the intensity of these fluxes could lead to changes in the salt balance of the Gulf. Incorporation of these recent physical advances into the chemical models should improve their quality. The potential implications for the interpretation of existing data and the design of new chemical research on the Gulf are discussed in Chapter X.

Geologic Setting

Loring and Nota (1973) have discussed the morphology, sediments, and surrounding geology of the Gulf of St. Lawrence in considerable detail. The Gulf of St. Lawrence region includes three major geologic provinces. The Appalachian region south of the St. Lawrence River and Gulf consists mostly of sedimentary rocks with some igneous outcrops and includes the Magdalen Shelf in the Gulf itself. The St. Lawrence Lowlands extend all the way from Lake Huron to the Strait of Belle Isle and include the areas of the Gulf to the north and east of the Laurentian Channel. The St. Lawrence River valley and the channels of the Gulf of St. Lawrence have been incised into the soft sedimentary rocks of the Lowlands. The Laurentian region borders the St. Lawrence

11

River and the Gulf to the north and is composed mostly of igneous and metamorphic rocks of Precambrian age. The diverse nature of these areas is reflected by differences in the chemistry of freshwaters entering the Gulf. For example, Loring and Nota (1973) reported relatively high concentrations of dissolved carbonate (≈ 1.4 mM) in discharge from the St. Lawrence River, which drains large areas of sedimentary rocks. In comparison, water from the Saguenay River, which drains the igneous and metamorphic rocks of the Laurentian region, is low in carbonate (≈ 0.1 mM).

The distribution of sediments in the Upper Estuary is controlled by the same physical processes responsible for the maintenance of the turbidity maximum. Fine-grained sediments, referred to as sandy (5-30 % by weight) pelites or muds are deposited only beneath the turbidity zone, along the northern side of the upper Estuary in a shallow depression (water depths less than 10 m) 15-20 km below the river mouth, and in the deeper eastern basin. Elsewhere, muddy (5-30 %) sands, very sandy (> 30 %) muds, sands, and gravels cover the floor (d'Anglejan and Brisebois 1974; Kranck 1979). In such a high energy environment with few permanent deposits of fine material, the source of the suspended matter for the turbidity maximum is puzzling. Silverberg and Sundby (1979) concluded that advection was more important than resuspension for the maintenance of the turbidity zone, and suggested that fine sediments found in the channel north of Île d'Orléans might be one source for the material. d'Anglejan (1981) stated that the intertidal mudflats, which are abundant on the south side of the Upper Estuary between Quebec City and Pointe aux Orignaux, are regions of temporary deposition and supply material to the turbidity maximum.

The Saguenay Fjord contains a variety of sediments (Loring 1975). Black, anoxic, sandy (5-10 %) muds are found in the upper shallow parts of the Fjord. These merge seaward into dark, greenish-grey muds that occupy the deep inner basin of the Fjord. Further downstream, very sandy (> 30 %) muds are found in the basins with muddy sands on the sills. Sand and gravel cover the floor at the mouth of the Fjord.

The grain size distribution of the sediments in the Lower Estuary and the open Gulf is mostly controlled by the submarine topography. Fine-grained sediments (95 % of the material < 0.05 mm in diameter), greenish-grey or calcareous (> 5 % $CaCO_3$) muds, occur in the deep central parts of the channels and some shelf valleys. Sandy muds (or their calcareous equivalents) cover the sides and lower slopes of the channels, grading into very sandy muds on the upper slopes of the principal channels, where they coexist with local outcrops of glacial drift. In the northern Gulf, the shelf sediments are mainly coarse to medium grained sands containing gravel and mud, or gravels. Limestone and dolomite gravels are found near Anticosti Island and on the outer Quebec/Labrador shelf. The better sorted sediments found on the Magdalen Shelf include gravels and coarse to fine sands containing some gravel and mud. Grain sizes decrease with increasing water depth and distance from shore or the tops of banks. The sandy to very sandy muds found in the shelf valleys are quite distinct from those found in the major channels of the Gulf.

The Gulf of St. Lawrence as a Setting for Chemical Oceanographic Research

Apart from its intrinsic interest as a major marginal sea of considerable economic and social significance, the Gulf of St. Lawrence has a number of features that make it a particularly interesting and rewarding area for chemical oceanographic research. The Gulf is almost surrounded by land, with the most important freshwater inputs concentrated at the western end of the Gulf and the only connections with the Atlantic Ocean at the eastern extremes of the system. This semi-enclosed nature and the separation between inputs and outputs greatly facilitates the study of both internal processes and exchanges with adjacent waters. Extensive studies of the physical oceanography of the region have characterized the principal features of the circulation and water mass structure of the Gulf. These are important considerations in the study of chemical processes, in the estimation of chemical fluxes and in the construction of budgets for chemical species. Understanding the flux of riverborne materials, both natural and anthropogenic, through the coastal zone is essential to the understanding of the impact of such materials on the ocean. The Gulf of St. Lawrence provides a rare natural laboratory for the study of such processes.

The St. Lawrence Estuary is a very large Estuary receiving the discharge from one of the world's major rivers and, like the Gulf itself, is geographically well defined. It displays behaviours ranging from that of a well mixed to that of a stratified estuary, and the Saguenay Fjord provides yet another type of estuarine environment in the same system. The St. Lawrence and the Saguenay rivers drain areas of very high population concentration and heavy industry respectively, making the estuaries useful for the study of the fate of contaminants at the fresh/saltwater boundary. Finally, the very high sedimentation rates and anoxic conditions found in surface sediments at the head of the Saguenay Fjord have produced an environment which has recorded signals indicative of climate and pollution history of the recent past in remarkable detail.

References

BEWERS, J. M. AND P. A. YEATS. 1979. The behavior of trace metals in estuaries of the St. Lawrence basin. Nat. Can. (Que.) 106: 149-161.

BUGDEN, G. L. 1981. Salt and heat budgets for the Gulf of St. Lawrence. Can. J. Fish. Aquat. Sci. 38: 1153-1167.

BUGDEN, G. L., B. T. HARGRAVE, M. M. SINCLAIR, C. L. TANG, J-C. THERRIAULT, AND P. A. YEATS. 1982. Freshwater runoff effects in the marine environment: The Gulf of St. Lawrence example. Can. Tech. Rep. Fish. Aquat. Sci. 1078: 89 p.

CANADIAN NATIONAL COMMITTEE. 1972. Discharge of selected rivers in Canada. Information Canada, Ottawa, Ont. 338 p.

D'ANGLEJAN, B.F. 1981. On the advection of turbidity in the St. Lawrence middle estuary. Estuaries 4: 2-15.

D'ANGLEJAN, B.F., AND M. BRISEBOIS. 1974. First subbottom acoustic reflector and thickness of recent sediments in the upper estuary of the St. Lawrence River. Can. J. Earth Sci. 11: 232-245.

DEP. OF FISHERIES AND OCEANS, CANADA. 1987. Canadian Fisheries Annual Statistical Review 1984. Econ. Commer. Anal. Ser. Surv. Stat. Rep. #1, vol. 17: 185 p.

DICKIE, L.M., AND R.W. TRITES. 1983. The Gulf of St. Lawrence, p. 403-425. In B. H. Ketchum [ed.] Ecosystems of the World 26. Estuaries and Enclosed Seas. Elsevier, Amsterdam.

DYER, K. R. 1973. Estuaries: A physical introduction. John Wiley and Sons, London. 140 p.

EL-SABH, M. I. 1975. Transport and currents in the Gulf of St. Lawrence. Bedford Inst. Oceanogr. Rep. BI-R-75-9: 180 p.

  1977. Oceanographic features, currents, and transport in Cabot Strait. J. Fish. Res. Board Can. 34: 516-528.

  1979. The lower St. Lawrence estuary as a physical oceanographic system. Nat. Can. (Que.) 106: 55-73.

FAIRBANKS, R. G. 1982. The origin of continental shelf and slope water in the New York Bight and the Gulf of Maine: Evidence from $H_2^{18}O/H_2^{16}O$ ratio measurements. J. Geophys. Res. 87: 5796-5808.

FORRESTER, W. D., AND P. E. VANDALL JR. 1968. Ice volumes in the Gulf of St. Lawrence. Bedford Inst. Oceanogr. Rep. 68-7: 16 p.

GREISMAN, P., AND G. INGRAM. 1977. Nutrient distribution in the St. Lawrence estuary. J. Fish. Res. Board Can. 34: 2117-2123.

JORDAN, F. 1973. The St. Lawrence system run-off estimates. Bedford Inst. Oceanogr. Data Rep. BI-D-73-10: 11 p.

KRANCK, K. 1979. Dynamics and distribution of suspended particulate matter in the St. Lawrence estuary. Nat. Can. (Que.) 106: 163-173.

LORING, D. H. 1975. Mercury in the sediments of the Gulf of St. Lawrence. Can. J. Earth Sci. 12: 1219-1237.

LORING, D. H. AND D. J. G. Nota. 1973. Morphology and sediments of the Gulf of St. Lawrence. Bull. Fish. Res. Board Can. 182: 147 p.

MATHESON, K. M. 1967. The meteorological effect on ice in the Gulf of St. Lawrence. McGill Univ. Mar. Sci. Cent. Ms. Rep. 3: 110 p.

NEU, H. J. A. 1982a. Man-made storage of water resources — a liability to the ocean environment? Part I. Mar. Pollut. Bull. 13: 7-12.

1982b. Man-made storage of water resources — a liability to the ocean environment? Part II. Mar. Pollut. Bull. 13: 44-47.

PETRIE, B., B. TOULANY, AND C. GARRETT. 1988. The transport of water, heat and salt through the Strait of Belle Isle. Atmos. Ocean 26: 234-251.

POCKLINGTON, R. 1982. Carbon transport in major world rivers: the St. Lawrence, Canada. Mitt. Geol.-Palaont. Inst. Univ. Hamburg 52: 347-353.

POCKLINGTON, R., AND F. C. TAN. 1987. Seasonal and annual variations in the organic matter contributed by the St. Lawrence River to the Gulf of St. Lawrence. Geochim. Cosmochim. Acta 51: 2579-2586.

POSTMA, H. 1967. Sediment transport and sedimentation in the estuarine environment, p. 158-179. In G. H. Lauff [ed.] Estuaries. Am. Assoc. Adv. Sci., Washington, DC.

REID, S. J. 1977. Circulation and mixing in the St. Lawrence Estuary near Islet Rouge. Bedford Inst. Oceanogr. Rep. BI-R-77-1: 36 p.

SEIBERT, G. H., R. W. TRITES, AND S. J. REID. 1979. Deepwater exchange processes in the Saguenay Fjord. J. Fish. Res. Board Can. 36: 42-53.

SILVERBERG, N., AND B. SUNDBY. 1979. Observations in the turbidity maximum of the St. Lawrence estuary. Can. J. Earth Sci. 16: 939-950.

SINCLAIR, M., G. L. BUGDEN, C. L. TANG, J.-C. THERRIAULT, AND P. A. YEATS. 1986. Assessment of effects of freshwater runoff variability on fisheries production in coastal waters, p. 139-160. In S. Skreslet [ed.] The role of freshwater outflow in coastal marine ecosystems. Springer-Verlag, Berlin.

STEVEN, D. M. 1974. Primary and secondary production in the Gulf of St. Lawrence. McGill Univ. Mar. Sci. Cent. MS Rep. 26: 116 p.

THERRIAULT, J.-C, AND G. LACROIX. 1975. Penetration of the deep layer of the Saguenay Fjord by surface waters of the St. Lawrence Estuary. J. Fish. Res. Board Can. 32: 2373-2377.

TRITES, R. W. 1972. The Gulf of St. Lawrence from a pollution viewpoint, p. 59-72. In M. Ruivo [ed.] Marine pollution and sea life. Fishing News (Books). London.

CHAPTER II

Distribution and Transport of Suspended Particulate Matter

P. A. Yeats

Marine Chemistry Division, Physical and Chemical Sciences Branch,
Department of Fisheries and Oceans, Bedford Institute of Oceanography,
P.O. Box 1006, Dartmouth, N.S. B2Y 4A2

Introduction

The nature, distribution, and transport of suspended particulate matter (SPM) in the St. Lawrence Estuary and the Gulf of St. Lawrence to some extent control the transport and behaviour of the chemical components in this system. Many inorganic and organic constituents undergo exchanges between dissolved and particulate phases depending upon their affinity for particulate material. While dissolved constituents that have very low affinity for particulate material may be regarded as conservative in water and are therefore distributed entirely as a consequence of water mixing processes, most chemical species have significant affinities for either organic or inorganic particles and are therefore subject to non-conservative transport that is greatly affected by the concentrations and types of particles in the system and their fate. Because of the control that SPM exerts on the behaviour of other water constituents, it is discussed prior to the other chemical components.

Upper St. Lawrence Estuary

The distribution, transport, and composition of SPM in the Upper St. Lawrence Estuary was first studied by d'Anglejan and Smith (1973). Their description of the SPM distribution is based on measurements over one complete tidal cycle at each of seven stations occupied in the summer of 1971. High concentrations were observed between Cap Maillard and Pte. aux Pic with a sharp reduction in the SPM levels from ≈ 20 mg·L^{-1} west of Pte. aux Pic to ≈ 5 mg·L^{-1} further downstream. The observed SPM concentrations in the area near the mouth of the Upper Estuary were only 1-2 mg·L^{-1}. Cross-estuary gradients were observed with lower concentrations towards the north shore. The changing location and intensity of the turbidity front near Pte. aux Pic was studied by repeated transmissometer transects. Tidal cycle observations at the stations upstream of Pte. aux Pic indicated that peak SPM concentrations at all depths occur near half-tide and that concentrations generally increased with depth. In the deeper water farther downstream, maximum SPM concentrations occur near high water.

Measurements of size distribution showed a remarkable spatial uniformity in the size distribution with a mean particle size of 5-7 μm. Mineralogical and chemical analysis of the fine surface water particulates indicated that, although the mineralogy is quite variable, chlorite and illite are the dominant clay minerals. Organic content of the SPM (determined by ashing) was quite low (5-40 % by weight).

An important contribution of this paper (d'Anglejan and Smith 1973) was the description of the estuarine turbidity maximum between Cap Maillard and Pte. aux Pic. The results suggested that the turbidity maximum is a steady-state feature resulting from a combination of two processes. The net non-tidal circulation carries particles downstream in the surface layer and upstream near the bottom, trapping those particles that are too large to be exported from the Estuary in the surface layer but too small to be permanently deposited. Particles in this narrow size range have a relatively long residence time in the turbidity maximum, which may explain the low organic content of the SPM. There is also extensive dissipation of tidal energy in the shallow portions of the turbidity maximum (see Chapter I), with the resulting resuspension adding material to the suspended load.

Based on a very limited number of stations, d'Anglejan and Smith (1973) described the turbidity maximum as the region of the Upper Estuary with high SPM concentrations (> 10 mg·L^{-1}) extending from the river downstream ≈ 100 km to the area near Pte. aux Orignaux. Downstream of Pte. aux Orignaux, high turbidity was only found on the southern side of the Estuary.

A number of subsequent studies have contributed data that can be used to better define the geographic extent of the turbidity maximum zone (d'Anglejan and Ingram 1976; Bewers and Yeats 1978; Silverberg and Sundby 1979; Kranck 1979; d'Anglejan 1981). The upstream limit of the turbidity maximum occurs between the centre of Île d'Orléans and its northern tip with the location of this limit being relatively insensitive to changes in river discharge (Silverberg and Sundby 1979). Although the landward limit of salt penetration is at approximately the same location as the upstream limit of the turbidity maximum, they do not always coincide. The highest turbidity is observed as a distinct SPM peak in the upstream part of the turbidity maximum zone between Île d'Orléans and Île aux Coudres. Downstream from this peak, SPM concentrations decrease quite rapidly although other more isolated regions of high SPM concentration are often seen, especially near the bottom. The downstream limit is located in the region of Pte. aux Orignaux, and is characterized by a distinct turbidity front (d'Anglejan 1981). The front runs diagonally across the Estuary, extending farther downstream on the southern side.

There has been a continued interest in the dynamics of the SPM in the Upper Estuary, particularly in the estuarine turbidity maximum, by a number of investigators (d'Anglejan and Ingram 1976; Silverberg and Sundby 1979; Kranck 1979; and d'Anglejan 1981).

In the deeper part of the Upper Estuary downstream of Île aux Coudres SPM concentration maxima are found at mid-depth and maximum concentrations throughout the water column occur 1 to 2 hours after low water (d'Anglejan and Ingram 1976). The SPM concentrations at any depth vary considerably with a semi-diurnal period. These features are predominantly determined by advective processes rather than resuspension. The main sources of the SPM in the mid-depth maxima are advection of resuspended sediment from the south shore or of SPM from the shallower regions of the Estuary farther upstream.

Another study (Silverberg and Sundby 1979) of tidal stations in the shallower, upstream part of the turbidity maximum revealed a similar dependence of SPM concentration on tide with highest concentrations again observed during the late ebb stages (i.e. shortly after low water — see Chapter I). In this study the maximum concentrations were generally observed at, or near, the bottom. Comparison of samples collected in May with those collected in November shows that the location and character of the turbidity maximum is not greatly affected by fairly large changes in river flow and location of the salt wedge. As did previous workers, Silverberg and Sundby concluded that the turbidity maximum is maintained by tidally forced, density-driven circulation combined with sediment resuspension. Since the sediments underlying the turbidity zone are predominantly sands and gravels, much of the resuspended sediment must be advected from adjacent areas such as the channel to the north of Île d'Orléans.

Data on the size spectra of SPM from the turbidity maximum region indicate a change from a fairly uniform distribution of all sizes in the river to a distribution increasingly comprising smaller-sized particles as the turbidity maximum is traversed (Kranck 1979; Silverberg and Sundby 1979). Within the turbidity maximum, intermediate sized particles ($\approx 10 \mu$m) dominate. The particle size distribution becomes better sorted and shows less temporal variability in the downstream portion of the turbidity maximum. Apparently, settling of the largest of the riverborne particles is accompanied by preferential resuspension/settling of intermediate size particles (4–16 μm) in the turbidity maximum. The outflow from the turbidity maximum is dominated by the smallest, most slowly settling particles.

16

Kranck (1979) discussed the importance of flocculation processes to the maintenance of the turbidity maximum. Although flocculation is not necessary for the formation of turbidity maxima, flocculation helps retain fine sediments in the Estuary and thus makes them available for resuspension. The apparent partitioning of organic and inorganic particulates, with preferential settling of organic matter in the turbidity maximum, is attributed to a greater tendency of organic matter to flocculate.

Some of the observations reported by Gobeil et al. (1981) on the major ion composition of the SPM in the river and turbidity zone are pertinent to a discussion of the SPM dynamics in the turbidity zone. Gobeil et al. observed increased Al, Si, Ca, Mg and Fe content and decreased Mn content in turbidity zone particles compared to those in the river water. Organic content, inferred by using these metal measurements to calculate the total inorganic content, decreased from $\approx 40\%$ in the river to $\approx 20\%$ in the turbidity maximum. The Si/Al, Fe/Al, and Mn/Al ratios decreased from the river to the turbidity maximum and Ca/Al and Mg/Al increased. Decreased Si/Al is consistent with the rapid removal of the coarsest fraction of the river suspended matter (i.e. sands with high Si content). Increased Al, Si, Ca, and Mg (and to a lesser extent Fe) contents in the turbidity maximum reflect the relatively greater removal of organic matter and the resuspension of predominantly inorganic sediments underlying, or adjacent to, the turbidity maximum zone. The fact that the Mn content decreases and the Mn/Al and Fe/Al ratios decrease could indicate that a significant fraction of the particulate Mn and Fe is associated with the organic matter. Flocculation of organo-iron colloids in estuaries has been described elsewhere (Sholkovitz et al. 1978). Resuspension of bottom sediments whose Mn and Fe contents had been reduced by diagenetic processes could also yield the observed results.

D'Anglejan (1981) reviewed information on the transport of turbidity in the Upper St. Lawrence Estuary. The turbidity maximum extends across the Estuary in the region of Île aux Grues. Along the south shore, a turbid plume flows from Île aux Grues downstream to the region off Pte. aux Orignaux. This is the main route for export of SPM from the turbidity maximum. Factors that influence the downstream advection of turbidity are the geometry of the south channel, tidal phase differences across the Estuary, resuspension of bottom sediments during periods of reduced stratification, and reinforcement of the plume by addition of highly turbid inshore waters, particularly from Baie Ste. Anne. The south channel plume ends in a distinct turbidity front in the Pte. aux Orignaux/Kamouraska area.

Saguenay Fjord

The distribution of SPM in the Saguenay Fjord has been studied gravimetrically by Sundby and Loring (1978), who surveyed the Saguenay both in May and September. In May, high suspended loads in the Saguenay's spring freshet led to concentrations between 10 and 15 mg·L^{-1} in surface water near the head of the Fjord, dropping to 2.5-5 mg·L^{-1} in surface water over the outer basin. Throughout the Fjord, concentrations > 1.0 mg·L^{-1} were only found in the upper 50 m of the water column. SPM concentrations in waters that are deeper, but not in close contact with the bottom, were generally < 0.25 mg·L^{-1}. Bottom waters show concentrations between 0.5 and 1.0 mg·L^{-1}, elevated with respect to other deep waters due to resuspension. The energy for resuspension is supplied by internal tides and density currents (see Chapter I). This three-layer structure in the SPM distribution was also seen in September, but concentrations are lower in all parts of the Fjord. In the surface layer maximum concentrations were between 2 and 3 mg·L^{-1}, in the deep layer concentrations were < 0.25 mg·L^{-1}, and in the bottom layer concentrations were between 0.25 and 0.5 mg·L^{-1}.

Chanut and Poulet (1979) studied SPM distributions and size spectra in the Saguenay on the same two cruises, using Coulter counter techniques. They found a two layer distribution of size spectra in May and a three layer distribution in September. In addition, Chanut and Poulet compared May SPM concentrations and size spectra in the Saguenay with those found in two sections across the St. Lawrence Estuary, one each side of the mouth of the Saguenay. Surface concentrations/spectra in the Estuary were very similar to those found near the mouth of the Saguenay Fjord and indicate the importance of advective mechanisms of particle transport.

Gulf of St. Lawrence

The first study of SPM in the Gulf of St. Lawrence was made by d'Anglejan (1969). He measured total and inorganic SPM concentrations in sections across Cabot Strait and the Gaspe passage on two cruises in August 1967 and spring 1968 as well as at a number of other stations throughout the Gulf. The observed concentrations were lower than expected for inshore water, generally being between 0.1 and 0.3 mg·L^{-1}. The Cabot Strait section exhibits a vertical concentration gradient. In the top 150 m the total concentrations were 0.3 to 0.4 mg·L^{-1} with only 10-15% inorganic. Between 150 and 350 m the total concentrations were similar but here the inorganic fraction made up 25-30% of the total. Below 350 m a bottom nepheloid layer was found with total concentrations increasing to 0.7 mg·L^{-1} with 45% inorganic. Similar vertical gradients were observed elsewhere in the Gulf. D'Anglejan and Smith (1973) also observed a bottom nepheloid layer extending $\approx$ 100 m off the bottom in the Lower St. Lawrence Estuary.

In April/May 1973, Sundby (1974) resurveyed the Gulf using more sensitive techniques. He was able to better characterize the three-layer distribution pattern (Fig. II.1 after Sundby 1974) originally seen by d'Anglejan. In general, Sundby found that the surface layer (depth < 20 m) contained SPM concentrations from 0.1 to 2.9 mg·L^{-1} with highest concentrations in the low salinity outflow from the St. Lawrence Estuary. The bulk of the water column, from 50 m depth to 50 m off the bottom, contained low levels of SPM of between 0.05 and 0.1 mg·L^{-1} with little variation between Cabot Strait and Pte. des Monts. The bottom layer contained 0.1 to 0.4 mg·L^{-1} with the highest levels occurring in the St. Lawrence Estuary. Although Sundby's original sampling did not proceed west of Pte. des Monts, sampling in early 1974 (Yeats et al. 1979) showed that similar trends continued through the Lower Estuary although at intermediate depths concentrations do increase as the head of the Estuary is approached (Fig. II.2 — data from Yeats et al. 1979).

The surface layer distribution is closely related to the surface circulation. Highest concentrations are associated with low salinity water that flows seaward from the St. Lawrence River through the Lower Estuary and along the shores of the Gaspe in the Gaspe current. Lowest concentrations are associated with inputs of saline surface water to the Gulf from the North Atlantic. This general pattern of surface SPM concentrations will obviously be modified by biological processes. Phytoplankton blooms will result in localized maxima that are not related to surface salinity.

The particle rich bottom nepheloid layer is well developed in both the Laurentian Channel, where increased concentrations are observed towards the head of the channel, and in the Esquiman Channel. D'Anglejan (1969) suggested a westward transport of bottom layer material in the Laurentian Channel as part of the general estuarine circulation of the southern Gulf. Sundby (1974) emphasized the importance of erosion of particles from shelves and coastlines as a means of supplying SPM to the near-bottom waters and, ultimately, the sediments of the troughs.

Several studies have indicated the importance of biogenic material in terms of both the character of SPM and its settling behaviour (d'Anglejan 1969; Syvitski et al. 1983; Silverberg et al. 1985; Tan and Strain 1983). Syvitski et al. (1983) described a number of visual observations made from a submersible in which they catalogued the various types of biogenic particles present in the different water layers. Tan and Strain (1983) have used carbon isotope ratios to identify some of the sources of the POC in the Gulf. General POC distributions in the Gulf are described in Chapter IV. By combining this information with the information on SPM distributions presented in this chapter, the

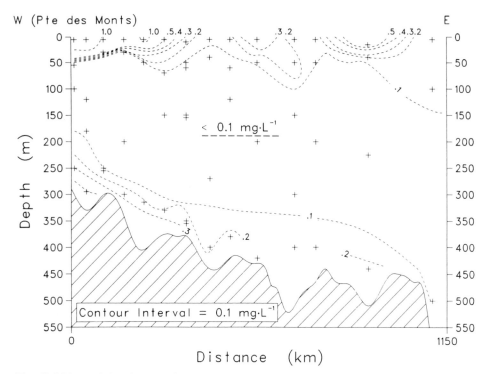

FIG. II.1 Vertical distribution of suspended particulate matter along the Laurentian Channel from Pte des Monts to the edge of the continental shelf outside Cabot Strait. Concentrations are expressed in mg·L^{-1}. (Redrawn from Sundby 1974 with permission from Can. J. Earth Sci.).

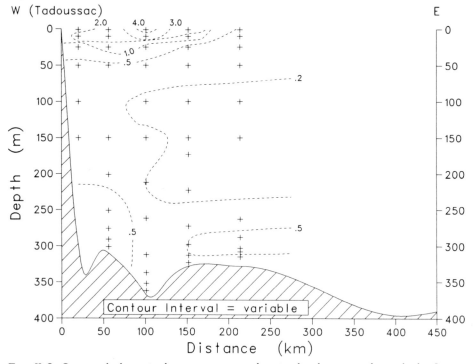

FIG. II.2. Suspended particulate matter on a longitudinal section through the Lower Estuary, May 1974.

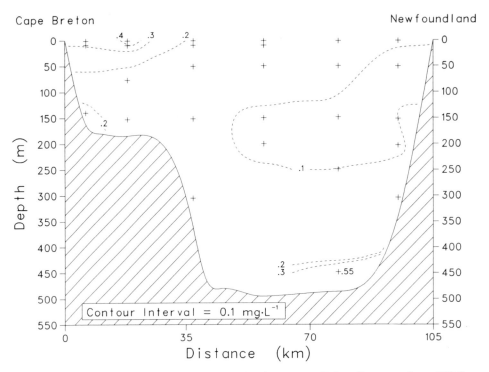

FIG. II.3a. Suspended particulate matter distribution in Cabot Strait in June 1975.

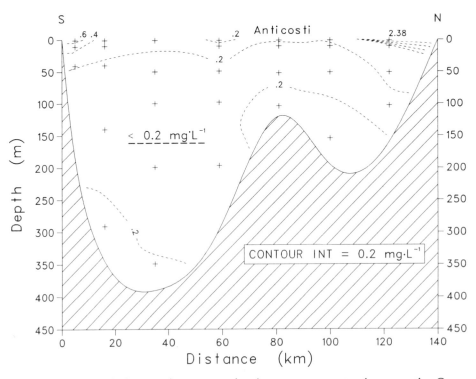

FIG. II.3b. Suspended particulate matter distribution on a section between the Gaspe Peninsula and the North Shore of Quebec in June 1975.

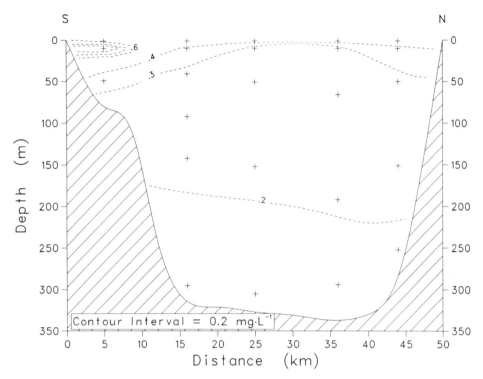

FIG. II.3c. Suspended particulate matter distribution on a transverse section at Pte des Monts in June 1975.

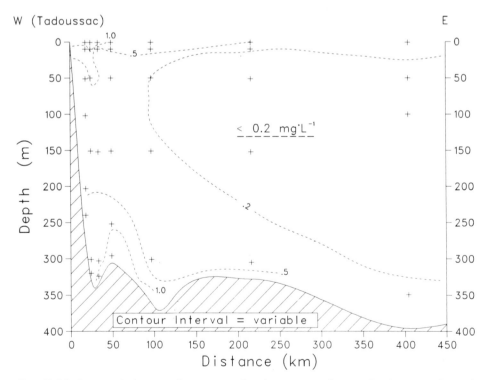

FIG. II.3d. Suspended particulate matter distribution on a longitudinal section through the Lower Estuary in June 1975.

percent organic content of the SPM can be estimated. Surface waters generally contain > 60 % organic matter except in low salinity surface waters affected by river discharge. Intermediate depth waters generally have 30-50 % organic matter and deep waters < 20 %. Several sections showing percent organic content of the SPM are shown in Chapter X. Silverberg et al. (1985) found that the organic carbon content of SPM caught in a Lower Estuary sediment trap increased gradually from ≈3.5 % in spring to ≈6 % in fall. The downward flux of organic matter, however, decreased by a factor of 3-4 over this time span because the total settling flux was much greater in the spring. The large fast settling particles that are caught by sediment traps are obviously much depleted in organic matter compared to average particles at surface or intermediates depths.

The most extensive study of SPM distributions in the Gulf of St. Lawrence is that of Sundby (1974). Because of ice conditions, however, he was unable to sample the southwest side of Cabot Strait. This is the location of the main outflow from the Gulf of St. Lawrence and is, therefore, an important area for study. In addition, his description of the SPM distribution is based solely on data acquired during one cruise in the spring. Studies of SPM in the Gulf of St. Lawrence were continued at BIO in the late 1970's with the result that better seasonal coverage of the SPM distribution at selected sections is now available. SPM distributions were measured at Cabot Strait and on several other sections in June (Cruise 75-015), July (76-021), August (79-024), and November (75-031). These sections are illustrated in Fig. II.3 to II.6.

This additional coverage confirms that the general distribution observed by Sundby (1974) persists at least through the summer months. The four Cabot Strait sections show that the SPM concentrations are higher in the shallow, predominantly outflowing water on the south side of the strait than in the deep (150-350 m), predominantly inflowing waters of the central and northern part of the strait. There is, however, considerable variability in the outflowing water concentrations being 0.07 mg·L^{-1} in July 1976, 0.11 mg·L^{-1} in November 1975, and 0.25 mg·L^{-1} in June 1975 and August 1979. Sundby estimated 0.2 to 0.4 mg·L^{-1} for May 1973 when ice prevented him from making very comprehensive measurements.

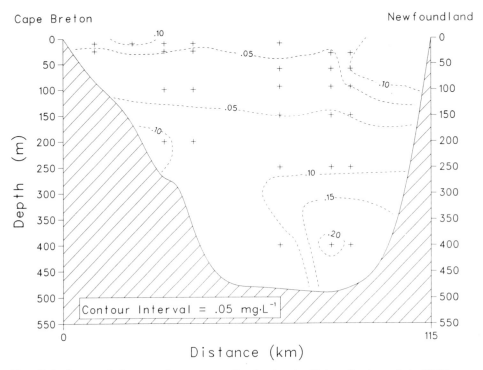

FIG. II.4. Suspended particulate matter distribution in Cabot Strait in July 1976.

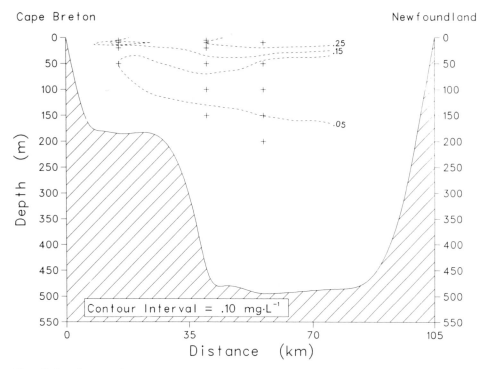

FIG. II.5a. Suspended particulate matter distribution in Cabot Strait in August 1979.

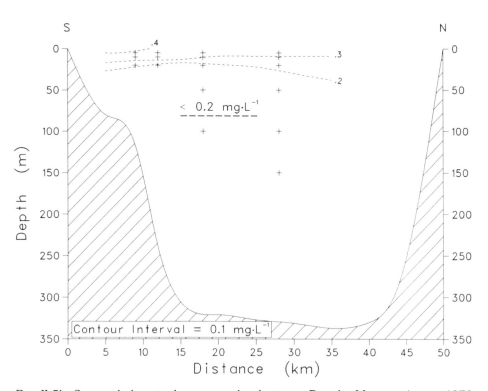

FIG. II.5b. Suspended particulate matter distribution at Pte. des Monts in August 1979.

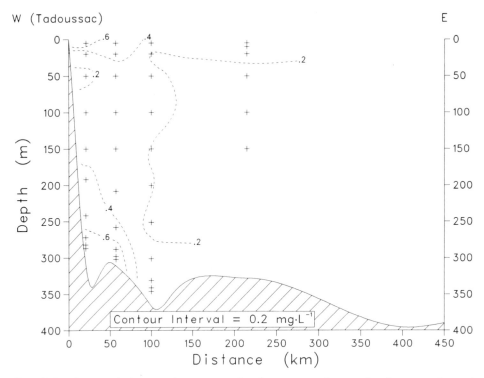

W (Tadoussac)   E

Contour Interval = 0.2 mg·L⁻¹

FIG. II.5c. Suspended particulate matter distribution on a longitudinal section through the Lower Estuary in August 1979.

In the region from the Gaspe passage to the head of the Laurentian Channel, concentrations increase in an upstream direction in all three layers. The close comparability of results from the various cruises would suggest that intermediate and deep layer concentrations are relatively invariant with time. The surface layer is more variable as would be expected with changing freshwater discharge and biological activity.

Sundby (1974) calculated a SPM budget for the Gulf based on the rather limited data available at that time. He came to the surprising conclusion that the net export of SPM at Cabot Strait equalled the river input and that all the sedimentation in the deep basins of the Gulf resulted from erosion of shallower regions. As a result of the additional SPM data collected since 1973, some useful flux estimates for nutrients and organic carbon discussed elsewhere in this book (Chapters III and IV), and more detailed estimates of water transport (El-Sabh 1977; Bugden 1981), an improved SPM budget can be calculated. This revised budget is illustrated in Fig. II.7. The main differences in the flux estimates in relation to those of Sundby (1974) occur at Cabot Strait. The revised budget uses water transports based on salt and mass conservation for three different seasons (spring, summer, winter) and corresponding SPM concentrations for these seasons based on the SPM data presented in this chapter and the water transport data in El-Sabh (1977). Inward and outward transports of SPM in various depth ranges were calculated for the three seasons using El-Sabh's water transport estimates, modified as suggested by Bugden (1981). River inputs of SPM were estimated from 2 years of monthly measurements of SPM in the St. Lawrence River (Yeats and Bewers 1982). The atmospheric input number is taken from Sundby (1974). POM transports through Cabot Strait are calculated using the same water flow data and analogous procedures in Chapter IV. The POM flux from the river can be estimated from the data presented by Pocklington and Tan (1983). The inorganic particulate matter transports are the differences between the SPM and POM transports.

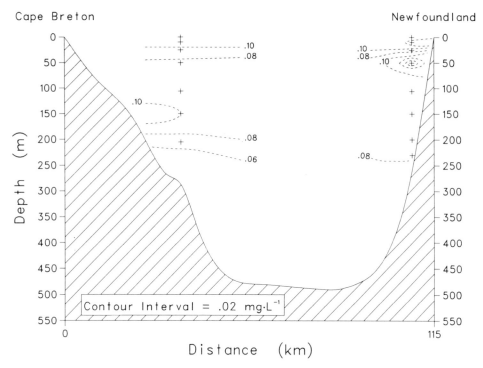

FIG. II.6a. Suspended particulate matter distribution in Cabot Strait in November 1975.

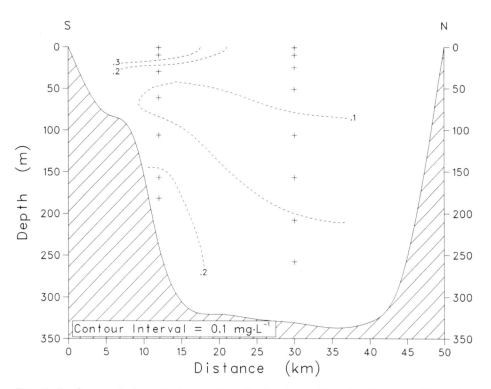

FIG. II.6b. Suspended particulate matter distribution at Pte. des Monts in November 1975.

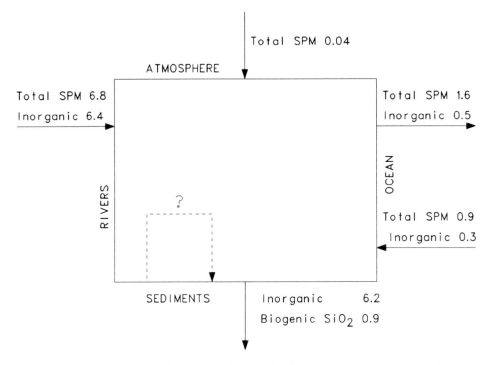

Total SPM 0.04

ATMOSPHERE

Total SPM 6.8
Inorganic 6.4

Total SPM 1.6
Inorganic 0.5

OCEAN

RIVERS

?

Total SPM 0.9
Inorganic 0.3

SEDIMENTS

Inorganic 6.2
Biogenic SiO₂ 0.9

FIG. II.7. Suspended particulate matter budget for the Gulf of St. Lawrence (all quantities in units of 10^6 t·yr^{-1}). The internally recycled sediment flux (dashed line) and the reasons for its uncertainty, are discussed in the text.

The budget shows that the SPM entering the Gulf in rivers is dominantly inorganic ($> 90\%$) but that which enters and leaves at Cabot Strait is approximately two thirds organic. This means that 6.2×10^6 t of inorganic SPM (almost completely derived from river input) is removed to the sediments in the Gulf, corresponding to a loss of 92% of the inorganic input. There is an approximate balance between the quantity of particulate organic matter entering and leaving the Gulf in these three water flows but this is a fortuitous balance because of the large quantity of internally produced biogenic material.

Removal fluxes to the sediments are rather different from Sundby's model. Sundby found a balance between the river input of SPM and the net export at Cabot Strait and therefore no net sedimentation. The revision to the budget discussed here leads to a substantially different estimate of the amounts of material being deposited within the Gulf. First, since most of the SPM exported at Cabot Strait is internally generated organic matter, most of the inorganic SPM (6.2×10^6 t·yr^{-1}) is removed to the Gulf sediments. To this will be added the net organic sedimentation and the input of biogenic silica to the sediments. The organic sedimentation will be very small ($\approx 2 \times 10^5$ t·yr^{-1}) since much of the settling particulate organic matter is redissolved (organic carbon concentrations in surface sediments are usually $< 3\%$ — see Chapter IV). The biogenic SiO₂ sedimentation has been estimated from a nutrient budget calculation to be 9×10^5 t·yr^{-1} (see Chapter III). These three sediment fluxes together give a total new input to the Gulf sediments of 7×10^6 t·yr^{-1}. Sundby (1974) estimated from sediment accumulation data that the minimum gross sedimentation rate in the Gulf is $\approx 1 \times 10^7$ t·yr^{-1}. If this number is correct, at least 3×10^6 t·yr^{-1} of sedimentary material must be eroded from areas of the Gulf where sediment is not accumulating and redeposited in regions where sediment is accumulating. Silverberg et al. (1986) have recently measured sediment accumulation rates at seven locations along the Laurentian Channel from the head of the Channel to Cabot Strait. These measurements give sedimentation rates from 2 to 160 times the minimum average sedimentation rate used by Sundby (1974). It is difficult to use the sedimentation rates at these seven stations to estimate an average rate for the regions of the Gulf where sediment is accumulating because the stations were all in the centre of the Laurentian Channel where sedimentation rates would be greatest, but they do show that the gross sedimentation rate taken from Sundby (1974) is most likely too low. If this rate is increased, the amount of material internally redistributed would have to be increased correspondingly.

References

BEWERS, J. M., AND P. A. YEATS. 1978. Trace metals in the waters of a partially mixed estuary. Estuarine Coastal Mar. Sci. 7: 147-162.

BUGDEN, G. L. 1981. Salt and heat budgets for the Gulf of St. Lawrence. Can. J. Fish. Aquat. Sci. 38: 1153-1167.

CHANUT, J. P., AND S.A. POULET. 1979. Distribution des spectres de taille des particules en suspension dans le Fjord du Saguenay. Can. J. Earth Sci. 16: 240-249.

D'ANGLEJAN, B. F. 1969. Preliminary observations on suspended matter in the Gulf of St. Lawrence. Marit. Sediments 5: 15-18.

1981. On the advection of turbidity in the St. Lawrence middle estuary. Estuaries 4: 2-15.

D'ANGLEJAN, B. F., AND R. G. INGRAM. 1976. Time-depth variations in tidal flux of suspended matter in the St. Lawrence estuary. Estuarine Coastal Mar. Sci. 4: 401-416.

D'ANGLEJAN, B. F., AND E. C. SMITH. 1973. Distribution, transport and composition of suspended matter in the St. Lawrence Estuary. Can. J. Earth Sci. 10: 1380-1396.

EI-SABH, M. I. 1977. Oceanographic features, currents, and transport in Cabot Strait. J. Fish. Res. Board Can. 34: 516-528.

GOBEIL, C., B. SUNDBY, AND N. SILVERBERG. 1981. Factors influencing particulate matter geochemistry in the St. Lawrence estuary turbidity maximum. Mar. Chem. 10: 123-140.

KRANCK, K. 1979. Dynamics and distribution of suspended particulate matter in the St. Lawrence estuary. Nat. Can. (Que.) 106: 163-173.

POCKLINGTON, R., AND F. C. TAN. 1983. Organic carbon transport in the St. Lawrence River. Mitt. Geol.-Palaont. Inst. Univ. Hamburg 55: 243-251.

SHOLKOVITZ, E.R., E.A. BOYLE, AND N.B. PRICE. 1978. The removal of dissolved humic acids and iron during estuarine mixing. Earth Planet. Sci. Lett. 40: 130-136.

SILVERBERG, N., AND B. SUNDBY. 1979. Observations in the turbidity maximum of the St. Lawrence estuary. Can. J. Earth Sci. 16: 939-950.

SILVERBERG, N., H.M. EDENBORN, AND N. BELZILE. 1985. Sediment response to seasonal variations in organic matter input, p. 69-79. In A.C. Sigleo and A. Hattori [ed.] Marine and estuarine geochemistry. Lewis Publishers, Chelsea, Mi.

SILVERBERG, N., H.V. NGUYEN, G. DELIBRIAS, M. KOIDE, B. SUNDBY, Y. YOKOYAMA, AND R. CHESSELET. 1986. Radionuclide profiles, sedimentation rates, and bioturbation in modern sediments of the Laurentian Trough, Gulf of St. Lawrence. Oceanol. Acta 9: 285-290.

SUNDBY, B. 1974. Distribution and transport of suspended particulate matter in the Gulf of St. Lawrence. Can. J. Earth Sci. 11: 1517-1533.

SUNDBY, B., AND D. H. LORING. 1978. Geochemistry of suspended particulate matter in the Saguenay Fjord. Can. J. Earth Sci. 15: 1002-1011.

SYVITSKI, J. P. M., N. SILVERBERG, G. OUELLET, AND K. W. ASPREY. 1983. First observations of benthos and seston from a submersible in the Lower St. Lawrence Estuary. Geogr. Phys. Quatern. 37: 227-240.

TAN, F. C., AND P. M. STRAIN. 1983. Sources, sinks and distribution of organic carbon in the St. Lawrence Estuary, Canada. Geochim. Cosmochim. Acta 47, 125-132.

YEATS, P. A., AND J. M. BEWERS. 1982. Discharge of metals from the St. Lawrence River. Can. J. Earth Sci. 19: 982-992.

YEATS, P. A., B. SUNDBY, AND J.M. BEWERS. 1979. Manganese recycling in coastal waters. Mar. Chem. 8: 43-55.

CHAPTER III

Nutrients

P. A. Yeats

*Marine Chemistry Division, Physical and Chemical Sciences Branch,
Department of Fisheries and Oceans, Bedford Institute of Oceanography,
P.O. Box 1006, Dartmouth, N.S. B2Y 4A2*

Introduction

The Gulf of St. Lawrence is an important fisheries region and, because of the role nutrients play in primary biological production, it is important to understand the distribution, circulation, and chemistry of these compounds. Nitrate (NO_3^-), phosphate (predominantly HPO_4^{2-}), and silicate (predominantly $Si(OH)_4$) constitute the three primary nutrient salts in the ocean. A deficiency of these nutrients, particularly the first two, may restrict or stop the growth of primary-producing organisms (Lucas and Critch 1974). Although nitrate is the most abundant available nitrogen species in solution in the sea, nitrogen is also present as nitrite, ammonia, and, in small quantities, as organic nitrogen compounds. In temperate coastal waters the lowest levels of nutrients are found in surface water during the summer months after a period in which plankton growth has been very rapid. Plant and animal debris falling from surface water disintegrates and redissolves, releasing the nutrients as it settles. As a result, nutrient concentrations tend to build up in subsurface water where there are few plants to consume them. Typical nutrient concentrations in these subsurface coastal waters would be $5-15$ $\mu mol \cdot L^{-1}$ (μM) silicate, $5-20$ μM nitrate, and $0.7-1.5$ μM phosphate. The nitrate to phosphate ratios for these waters are generally considerably lower than the $15:1$ or $16:1$ ratio found in subsurface pelagic waters.

Rivers generally contain higher nutrient concentrations than surface ocean waters, so they can be an important source of nutrients to estuarine and nearshore waters. Estuaries are characteristically systems of high primary productivity, and in general much of the energy fixed during photosynthesis is exported in detrital form. Furthermore, estuarine circulation can involve considerable entrainment of subsurface nutrient-rich waters into the seaward flowing surface waters. Consequently, estuaries can greatly influence the productivity of the surrounding coastal waters. The Gulf of St. Lawrence is particularly noteworthy in this respect. Sutcliffe (1973) showed strong correlations between variability of freshwater input and fish production in the Gulf, which implies that changes in nutrient entrainment resulting from changes in freshwater discharge may be important. In addition, Neu (1975) has drawn attention to the potential harm in modulating the freshwater input to the Estuary with dams. The change in entrainment that results from the damming of rivers could have an adverse effect on biological processes in the Gulf. It is not clear whether an increase in freshwater flow would increase or decrease the rate at which nutrients are mixed into the surface water, nor how a change in the seasonality of the flow might interact with the cycling of nutrients within the Gulf. This complex problem has been considered by Sinclair et al. (1986 — see also Chapter I).

In this chapter, nutrient distributions in the St. Lawrence Estuary and Gulf are described using both published data and some previously unpublished data from BIO that is presented here for the first time. This distributional data is then used to develop a conceptual picture of the physical and chemical mechanisms responsible for the distributions. Next, the interelement relationships inherent in nutrient regeneration are described in more detail and finally nutrient budgets for the Gulf are reviewed.

Nutrient Distributions

A general description of the nutrient distributions in the Gulf of St. Lawrence can be developed from the McGill University International Biological Programme (IBP) Studies (Bulleid and Steven 1972; Steven et al. 1973) summarized by Steven (1974) and the discussion of the early nutrient surveys conducted by the Bedford Institute of Oceanography (Coote and Hiltz 1975; Coote and Yeats 1979). The main features of the summer nutrient distributions in the Gulf are illustrated by a transect along the Laurentian Channel from the Lower St. Lawrence Estuary to Cabot Strait (Fig. III.1 after Coote and Yeats 1979). This figure shows the general increases with depth of all three nutrients that are observed throughout the Gulf (except the Saguenay Fjord) and the increases in nutrient concentrations in the deep water between Cabot Strait and the Estuary. Surface concentrations are generally quite low, except in the Estuary where input from the St. Lawrence River and entrainment of subsurface waters add nutrients to the surface layer. In winter (February, 1973), higher surface concentrations were found throughout the Gulf but deep water concentrations were unchanged from the summer. Coote and Yeats (1979) also measured nutrient concentrations at several sections across the Laurentian Channel and on a section from the Magdalen Shelf to the Strait of Belle Isle. These sections all showed the same general features as those depicted for the Laurentian Channel transect.

The IBP surveys concentrated on the surface and intermediate layers (Steven 1974). Stations on a very extensive grid throughout the Gulf were occupied on a number of occasions throughout the spring and summer months. In April, intermediate concentrations of all three nutrients were observed in the ice-free part of the Gulf. Higher concentrations of Si and N were found in the Lower Estuary and east of the Gaspe

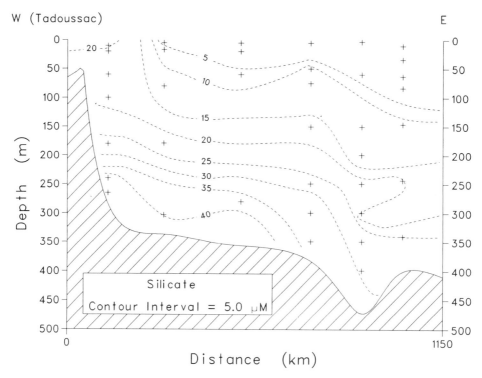

FIG. III.1. Summer nutrient distributions in the Laurentian Channel. Concentrations are in micromoles per litre. (For the locations of these and other sections, see Fig. I.1 and I.3.). (Redrawn from Coote and Yeats 1979).

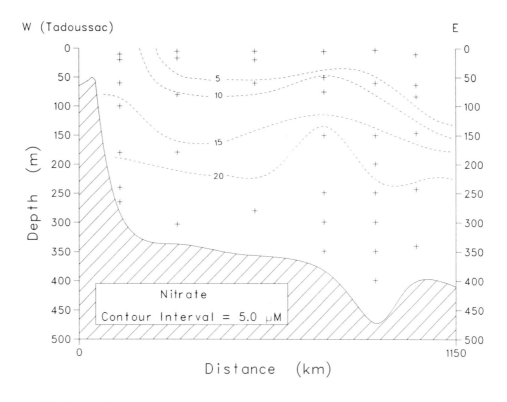

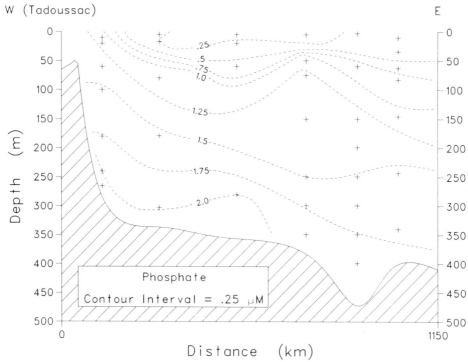

FIG. III.1. (*Concluded.*)

Peninsula, but elevated P levels were restricted to the latter area. By May, elevated nutrient levels were found only in the area to the west of Anticosti Island and around the end of the Gaspe Peninsula. The period from June to September showed the spread of higher Si and N concentrations eastward into the central Gulf, but P remained low except in isolated patches in the Lower Estuary and around the Gaspe peninsula. The northeastern portion of the Gulf, as defined by a line joining the middle of Cabot Strait and the east end of Anticosti Island, remained low in all three nutrients through September. In general, concentrations of all three nutrients decreased from maxima in the upper reaches of the Lower Estuary to lower concentrations in the eastern and northern Gulf; there is also a general decrease from highest levels in all regions in April to lowest levels in September.

Steven was the first to document the increasing concentrations of nutrients at intermediate depths between Cabot Strait and the Lower Estuary and to link these increases with nutrient regeneration processes. He also observed high intermediate-depth nutrient concentrations in the region between Pointe des Monts and the Gaspe Passage. The doming of nutrient, temperature, and salinity isopleths in this region has been shown subsequently to be a characteristic of the cyclonic circulation in the Anticosti Gyre that is a feature of this area (Sevigny et al. 1979). Doming of nutrient isopleths in this area is also evident in the Laurentian Channel transects of Coote and Yeats (1979).

Nutrient distributions in the Upper St. Lawrence Estuary were first discussed by Subramanian and d'Anglejan (1976) (silicate) and Greisman and Ingram (1977) (nitrate). In both these studies, essentially conservative mixing was observed. Small deviations from linear salinity–nitrate relationships were attributed to nutrient consumption, variability in the freshwater source, and amplitude of the internal tide.

The results of the nutrient analyses from the Upper Estuary on two BIO cruises in May and September, 1974, are illustrated by distributional plots of silicate, nitrate, and phosphate on the axial section along the main navigational channel of the Estuary (Fig. III.2 and III.3). No major transverse variability was observed on these cruises.

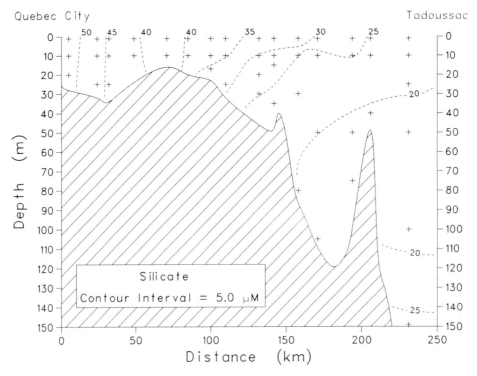

FIG. III.2. Nutrient distributions in the Upper St. Lawrence Estuary, May 1974.

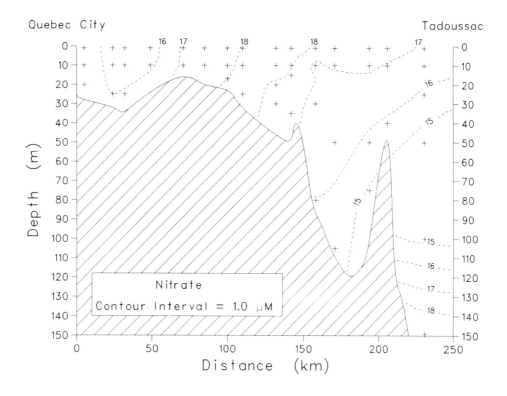

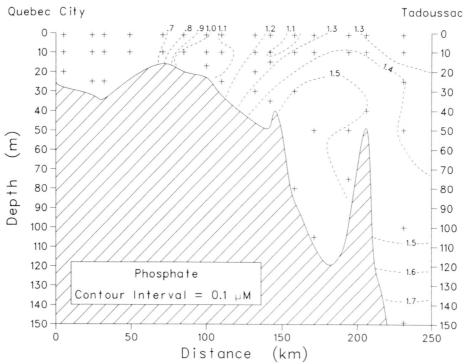

FIG. III.2. (*Concluded.*)

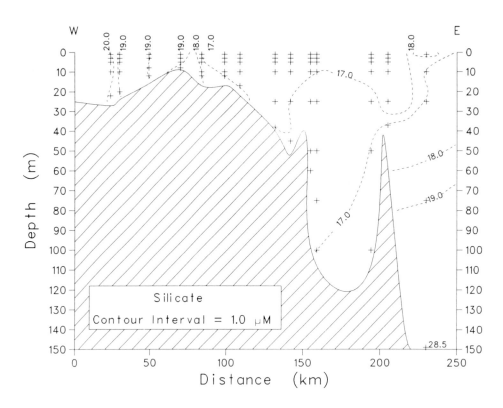

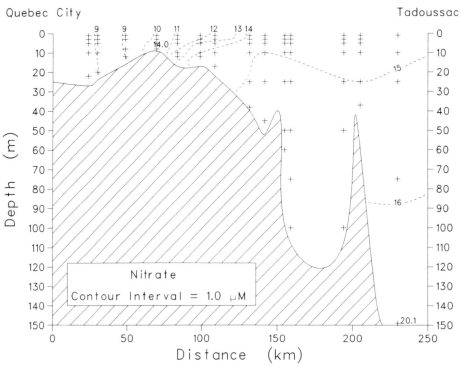

FIG. III.3. Nutrient distributions in the Upper St. Lawrence Estuary, September 1974.

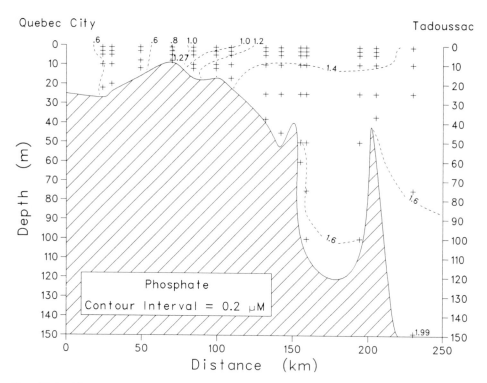

FIG. III.3. (*Concluded.*)

Downstream of the limit of salt intrusion in May, 1974, the silicate distribution simply reflects conservative mixing as illustrated by the plot of silicate versus salinity (Fig. III.4). The concentrations for samples from stations 1, 2, and 5 are all elevated with respect to the mixing relationship. Stations 1 and 2 are in freshwater several kilometres upstream of the limit of salt intrusion and the mechanism responsible for the decrease in concentration from stations 1 and 2 to stations 3 and 4 cannot be related to estuarine mixing processes. The elevated concentrations on station 5 probably result from inputs of high silicate freshwater through the channel to the north of Île d'Orléans.

The observed behaviour of silicate on stations 1 to 4 in the freshwater regime is puzzling. The progressive decrease in silicate concentration in the downstream direction is independent of salinity and completely at odds with the conservative relationship between silicate and salinity in the Estuary. The salt content on stations 1 to 4 varied between 40 and 110 ppm with no obvious trend towards increasing salinity in the downstream direction. The observed variability in the freshwater regime could reflect temporal changes in the riverine silicate concentrations; however, in that case, a linear relationship between silicate and salinity would not be expected in the estuarine regime. Nitrate and phosphate concentrations on these stations are virtually constant and lower than in the estuarine region. Interestingly, the behaviour of fluoride follows that of silicate, concentrations in the freshwater regime exceed those at the zero salinity intercept of an otherwise linear fluoride versus salinity relationship (Young 1976). If similar variability in the silicate concentration upstream of the limit of salt intrusion occurs in other rivers, this feature could lead to the impression of non-conservative mixing unless the zero salinity intercept of the mixing curve is carefully established.

In September, the silicate concentration in the Estuary is constant at 17.0 μM (Fig. III.5). In the freshwater regime, concentrations increase upstream as they did in May. As was the case in May, there is no evidence for nonconservative mixing within the Estuary. The distributions of nitrate and phosphate are considerably different from those of silicate. Both nitrate and phosphate have uniformly low concentrations in the freshwater regime and show significant increases in concentration coincident with the upstream limit of salt intrusion.

The concentration of nitrate in May, 1974 (Fig. III.2) increased to maximum values in the downstream portion of the estuarine turbidity maximum and then decreased farther downstream. The nitrate versus salinity relationship (Fig. III.4) shows maximum nitrate concentrations in the 5 to 20 salinity range and lower concentrations at both lower and higher salinities. This type of curved mixing relationship is indicative of produc-

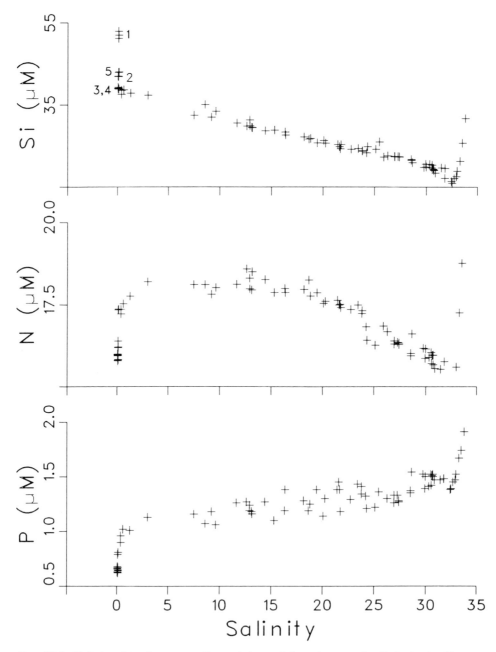

FIG. III.4. Relationships between silicate/nitrate/phosphate and salinity in the Upper St. Lawrence Estuary, May 1974. Some low salinity data points are labelled with station numbers.

tion of dissolved nitrate within the Estuary. A similar nitrate versus salinity relationship is seen in September (Fig. III.5) except that in this case the river concentration is approximately one half that in May. As a result, the curve does not show a maximum at intermediate salinity but a positive curvature is still evident.

The concentration of phosphate in May (Fig. III.2) increases from the freshwater stations to maximum values in the deep water of the eastern basin. The plot of phosphate versus salinity (Fig. III.4) shows a highly scattered, approximately linear, relationship with a small positive slope for samples from stations 5 to 17. The zero salinity intercept,

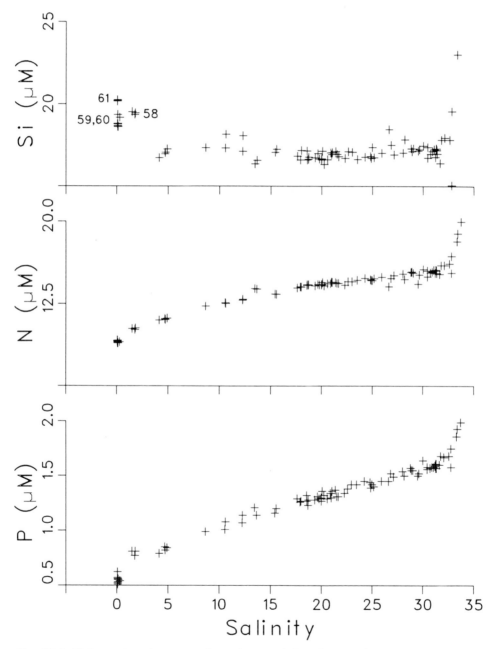

FIG. III.5. Relationships between silicate/nitrate/phosphate and salinity in the Upper St. Lawrence Estuary, September 1974. Some low salinity data points are labelled with station numbers.

however, is 50 % higher than the concentration actually observed in the freshwater regime. The concentrations of near surface samples from stations 25 to 28 at the head of the Lower Estuary are lower than those found at similar salinities on stations 1 to 17. The lower concentrations probably reflect biological uptake of phosphate in the Lower Estuary.

In September, phosphate concentrations at high salinity in the eastern basin were unchanged from the spring but, in the river and at intermediate salinity, concentrations were reduced compared to those in spring. An approximately linear, (except for the river samples) phosphate versus salinity relationship was again observed, but in this case the slope was greater and the zero salinity intercept was 30 % higher than the freshwater concentration. The scatter about the line was markedly reduced compared to the May data and the samples from the Lower Estuary were indistinguishable from the others. Although the scatter in the phosphate versus salinity relationships means that the exact forms of the relationships cannot be established, the similarity of the phosphate and nitrate distributions suggest that the phosphate versus salinity relationships are, like the nitrate ones, non-linear with slight positive curvature. The abrupt and fairly large increase in phosphate concentration at low salinity could result from the mortality of freshwater plankton incapable of surviving in a saline environment as proposed by Morris et al. (1978) for the Tamar Estuary. However, the character of the phosphate distributions, which generally show increased concentrations toward the bottom, and the coincidence of the increases with the turbidity maximum suggests that the predominant source of the additional phosphate is the sediments.

The concentrations of nitrate and silicate in the deep water of the eastern basin correspond closely to those at 30-50 m depth on stations in the Lower Estuary. Temperature and salinity are also very similar. It is evident that at these particular times the water in the eastern basin has its origin in the Lower Estuary and that the nitrate and silicate concentrations in the eastern basin simply reflect concentrations in the source water at sill depth (45 m). Phosphate concentrations on the May cruise are higher in the eastern basin (1.5 μM) than at 30-50 m depth in the Lower Estuary (1.4 μM). In the Lower Estuary, concentrations in excess of 1.5 μM are found only at depths greater than 100 m. The salinity at these depths is 33 whereas in the eastern basin the salinity of only one sample (31.8) exceeds 30.7. Several explanations for the high phosphate concentrations in the deep water of the eastern basin are possible. Phosphate may be released from the bottom sediments as suggested to explain the elevated levels farther upstream, mineralization of particulate organic phosphate entering the deep water of the eastern basin may be occurring, or the levels in the eastern basin may reflect concentrations near to the sill depth in the Lower Estuary during winter or earlier in the spring. Higher near surface nutrient levels are known to occur in the Gulf of St. Lawrence in winter (Coote and Yeats 1979). Temporal variations in the phosphate content of the water flowing into the eastern basin would also provide an explanation for the large scatter in the phosphate versus salinity relationship for May 1974.

Considerably elevated concentration of all three nutrients are observed for deep samples (salinity > 31) from stations in the Lower Estuary. These stations are at the head of the Laurentian Channel, the main deep channel in the Gulf of St. Lawrence. Increased concentrations in these samples are undoubtedly due to nutrient uptake and regeneration processes in the Gulf.

The deep water silicate concentrations in the Gulf of St. Lawrence are considerably greater than those at similar depths in the adjacent northwest Atlantic Ocean. It is also evident, from the data presented in Coote and Yeats (1979), that these elevated concentrations extend out into the Laurentian Channel east of Cabot Strait. In 1976, a cruise was conducted partly to investigate the extent of these elevated deep water concentrations. Two transects for all three nutrients from this cruise are illustrated in Fig. III.6, III.7, and III.8. The first transect runs due north along 55°W from 42°N in the northwest Atlantic to 45°20′N on the Newfoundland Shelf, and the second from 44°30′N, 55°W over the Newfoundland slope through the Laurentian Channel to Cabot Strait.

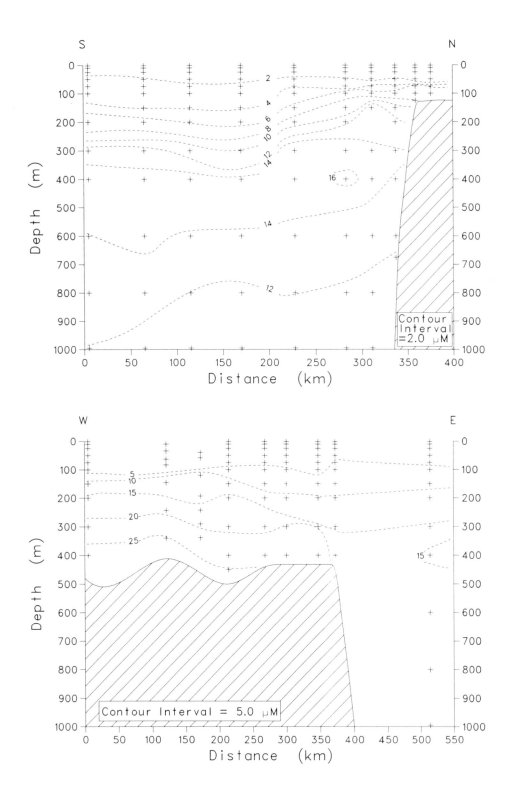

FIG. III.6. Silicate distribution in waters east of Cabot Strait. South to north section is along 55°W. West to east section is along the axis of the Laurentian Channel from Cabot Strait to 55°W. (See Appendix, cruise 76-021, for exact station locations).

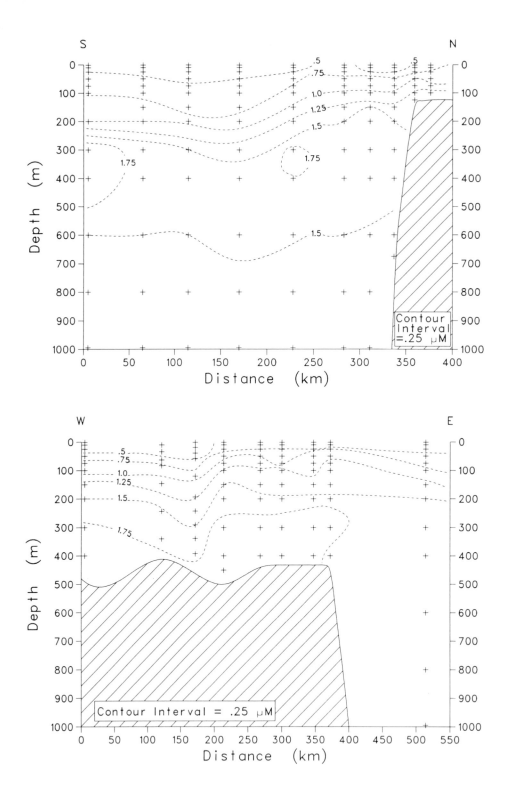

FIG. III.7. Phosphate distribution in waters east of Cabot Strait. South to north section is along 55°W. West to east section is along the axis of the Laurentian Channel from Cabot Strait to 55°W. (*See* Appendix, cruise 76-021, for exact station locations).

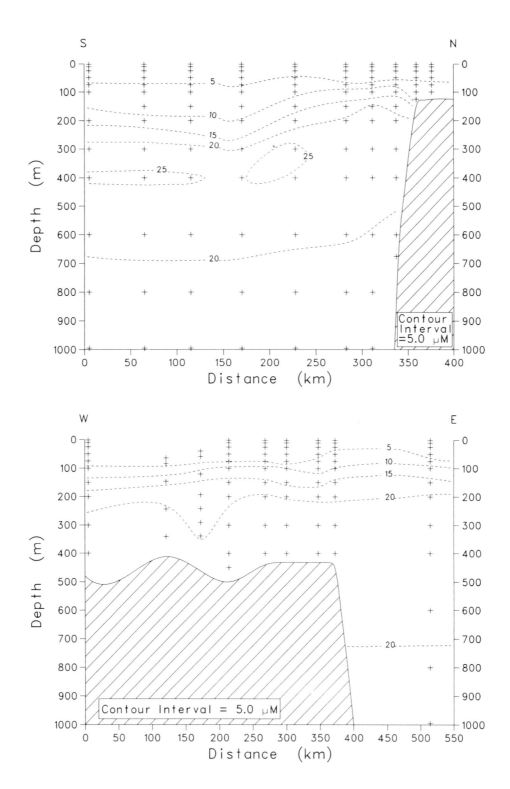

FIG. III.8. Nitrate distribution in waters east of Cabot Strait. South to north section is along 55°W. West to east section is along the axis of the Laurentian Channel from Cabot Strait to 55°W. (See Appendix, cruise 76-021, for exact station locations).

The north/south silicate section shows elevated concentrations between 200 and 500 m over the slope compared to similar depths in the deep water at 42°N. These concentrations, which extend into the bottom water on the edge of the Newfoundland shelf, are, however, considerably lower than those found throughout the deep water of the Laurentian Channel from Cabot Strait to the shelf break. In fact, there is a very marked horizontal front in silicate concentration between station 14 at the shelf break and station 17 only 25 km into the Laurentian Channel. From station 17 to station 32 in Cabot Strait, concentrations increase very gradually.

Phosphate behaves much like silicate except that the increase in concentration from the Atlantic to the Laurentian Channel is much smaller (Fig. III.7). Unlike silicate and phosphate, nitrate (Fig. III.8) shows no increase from the slope water to the Laurentian Channel, but concentrations at 100–300 m are somewhat elevated over the slope compared to levels at 42°N. In all three cases, the deep water concentrations in the Laurentian Channel outside Cabot Strait were temporally invariant over the 1971–76 period. Results from cruises in 1971 and 1972 (Coote and Yeats 1979) show virtually the same concentrations as the 1976 results.

Nutrient Cycling Mechanisms

Mechanisms for supply of nutrients to the surface waters of the Gulf of St. Lawrence were described by Steven (1974). In the open Gulf (east of 63°N plus all of the northern Gulf) surface nutrients are supplied by the vertical mixing and diffusion mechanisms active in all temperate seas. As a result, at the onset of spring, nutrients are distributed more or less uniformly throughout the top 100 m of the water column. These elevated surface concentrations are rapidly depleted during the spring phytoplankton bloom and remain low throughout the summer.

In the estuarine part of the Gulf, nutrients are supplied to the surface layer by vertical mixing and upwelling in the St. Lawrence Estuary (see Chapter I). These nutrients are then advected horizontally eastward by the Gaspe Current. Several mechanisms for surface water enrichment were proposed by Steven. The seaward flow of the St. Lawrence and Saguenay rivers generates sheer that entrains nutrient-rich subsurface water which mixes with the fresher water in the surface layer. Internal tides can also bring intermediate layer water to the surface. Finally, rivers contain elevated levels of nutrients that are supplied directly to estuarine surface waters.

The general vertical nutrient distribution in the Lower Estuary is described by Steven as having elevated nutrient concentrations in the top 25 m, minimum at 50–100 m and increasing concentrations at greater depths. This type of profile is predominant in the spring, but is also observed later in the year. No specific physical mechanism for this type of profile was identified by Steven, although one or more of the general mechanisms identified in the previous paragraph must play a role. Concentrations in the deep water were more extensively studied by Coote and Yeats (1979). They found a general increase in deep water concentrations from Cabot Strait to the head of the Laurentian Channel. Similar intermediate water trends were seen by Steven (1974). A general mechanism for nutrient cycling in the Gulf can be developed from these studies.

Subsurface water with high concentrations of nutrients flows inward (i.e. upstream from Cabot Strait); it is then mixed into the surface layer by entrainment with water from St. Lawrence River, by upwelling or by other mixing processes. This body of water then flows out toward Cabot Strait in the surface layer. In the surface layer nutrients are incorporated into plankton, some of which sink back into the intermediate and deep layers where the nutrients are regenerated to be returned to the surface farther upstream. Consequently, there is a cycling of nutrients within the Gulf between the subsurface and surface layers. Nutrients cycled in this way may spend some time in the surface sediments and a portion of the sedimentary fraction will become permanently buried. The entrainment of nutrients in the St. Lawrence Estuary is particularly important as the major source of nutrients to the highly productive surface waters of the Lower Estuary and Gaspe Current systems and also as a contributor of nutrients to the rest of the southern Gulf. The net effect of this cycling pattern is to trap nutrients within the Gulf of St. Lawrence and in the Laurentian Channel out to the edge of the continental shelf.

The physical mechanisms for the introduction of nutrients to the surface layer of the Estuary have been considered in more detail by Therriault and Lacroix (1976), Sinclair et al. (1976), Greisman and Ingram (1977), and Sevigny et al. (1979). Therriault and Lacroix (1976) described a mechanism whereby internal tides in the Lower Estuary can be responsible for penetration of nutrient rich intermediate waters from the Lower Estuary into the eastern basin of the Upper Estuary. Intense mixing processes in the eastern basin cause nutrient enrichment of the overlying waters and subsequent advection into Lower Estuary surface waters. Greisman and Ingram (1977) followed the estuarine mixing of nitrate from the St. Lawrence River and calculated the contribution of freshwater and Lower Estuary intermediate depth seawater to the surface mixture from Lac St. Pierre (100 km upstream of estuarine mixing) to Pointe des Monts. They found quite close agreement between calculated nutrient concentrations and those actually observed. These calculations show that in June 1975 river nitrate contributed 50 % of the surface water nitrate in the eastern basin of the Upper Estuary and slightly less than 25 % of the surface layer nitrate in the Lower Estuary.

The dynamics of circulation and mixing in the Lower Estuary and western Gulf of St. Lawrence are not well understood. At the head of the Lower Estuary, freshwater discharge from the St. Lawrence River will have a major effect on the dynamics. In addition, the rapid shoaling at the head of the Laurentian Channel may contribute to the injection of nutrients into the highly productive surface waters of the Lower Estuary. Farther seaward, wind-driven surface circulation will be increasingly important. A number of studies of the physical mechanisms for mixing in these regions have been done (see El-Sabh 1975, 1979; Ingram 1979) and the nutrient distributions will be governed to a large extent by the same processes. One study of the Anticosti Gyre does include a discussion of the nutrients. Sevigny et al. (1979) show that the cyclonic circulation in the gyre results in a general shoaling of the nutrient isolines as the centre of the gyre is approached; however, a strong shallow thermocline prevented nutrient enrichment of the surface layer during the period of the survey.

Nutrient Regeneration

An important component of nutrient cycling within the Gulf of St. Lawrence is the regeneration of nutrients within the intermediate and deep waters of the Gulf. The general vertical distribution of nitrate and phosphate shows that concentrations of both of these nutrients increase fairly rapidly from low levels in surface waters to high levels in the 75–150 m depth range. Between 150 m and the bottom, nitrate and phosphate concentrations continue to increase but much more gradually. Silicate concentrations, on the other hand, tend to increase much more uniformly with depth, generally showing little or no evidence of a change in gradient at intermediate depths. These trends are an indication of the tendency for nitrate and phosphate to be predominantly regenerated at shallow depths, and for silicate to be regenerated more uniformly throughout the water column and probably in the surface sediments as well.

Nitrate to phosphate ratios of 15–16:1 have been obtained consistently in subsurface water outside the continental shelf. A larger range has been observed in coastal regions usually with considerably lower N:P ratios. The N:P ratio is determined by the relative amounts of nitrate and phosphate that are utilized and released by micro-organisms, the N:P ratios of mixing water masses and external sources of nitrate and phosphate, such as rivers. The linear regression of phosphate on nitrate derived by combining all the results presented in Coote and Yeats (1979) for stations with depths > 200 m is:

$$N\ (\mu M) = 15.1\ P\ (\mu M) - 4.7 \qquad n = 329,\ r = 0.88$$

The N:P ratio is thus very similar to the traditional open ocean value but with a significantly negative intercept.

During the summer months, biological activity reduces the concentrations of all three nutrients in the top 50 m of the water column. Nitrate seems to be the limiting nutrient and in most of the Gulf the nitrate concentration is $< 0.2\ \mu M$ in the upper 25 m. Phosphate is less depleted with the result that the N:P ratio is often reduced to one or less. From the regressions of P on N, the phosphate concentration corresponding to [N] = 0 is $0.3\ \mu M$. In the February, 1973 cruise (Coote and Yeats 1979), surface depletion was not observed, and surface nitrate and phosphate concentrations fitted closely to the deepwater regression line.

The linear regressions referred to above are derived from large data sets, including data from surface water, water entering through Cabot Strait and the deep water of the troughs. The consistency in the circulation pattern and concentrations of nutrients and dissolved oxygen from year to year in all locations in the deep inflowing layer of the Laurentian Channel allowed Coote and Yeats (1979) to relate changes in the nutrient and oxygen concentrations along the channel to regeneration of nutrients solely in this deep layer. In doing this, it was assumed that little or no water from other sources was being incorporated into this layer. The fact that there was little change in salinity with distance at any particular depth supported this assumption. As the water flowed in the deep layer from Cabot Strait to the Estuary, the concentrations of silicate, phosphate, and nitrate increased and the oxygen concentration decreased as organic matter decayed in the deep water. The net changes in the concentrations of silicate, nitrate, phosphate, and oxygen at 250 m depth were 30, 5.6, 0.8, and $- 110\ \mu M$ (as O_2), respectively, on Cruise 71-027; and 25, 3.5, 0.50, and $- 75\ \mu M$, respectively, on Cruise 72-017. These findings are in agreement with those of Therriault and Lacroix (1976) who concluded from their study that high nutrient levels accompanied low oxygen concentrations. The concentration changes correspond to $\Delta O: \Delta N: \Delta P$ atomic ratios of 278:7.0:1 and 298:7.2:1 for nutrients regenerated in the deep layer from Cabot Strait to the Estuary. These O:P ratios were close to the accepted values of 276:1 (Redfield et al. 1963) but the N:P ratios, although virtually identical for the two cruises, were considerably lower than the ideal oceanic value of 16:1. However, low N:P ratios are not unusual in estuarine and coastal environments (Ketchum et al. 1958; Sen Gupta and Koroleff 1973). The low N:P ratio may indicate incomplete regeneration of nitrate compared to phosphate (which would include the conversion of organic N to ammonia rather than nitrate); alternatively, the organic matter being regenerated may be low in nitrogen. The ratio of Si:P from this analysis is $\approx 40:1$.

It is now possible using more recent data to repeat this calculation in a statistically more rigorous manner and to include carbon in the Redfield ratio estimate. Cruises in April–May 1973 (73-012 — See Appendix), July–August 1974 (74-028), and May–June 1975 (75-015) all included stations spaced along the Laurentian Channel from Cabot Strait to the Lower Estuary. Data for dissolved oxygen and the carbon isotope ratio of the dissolved inorganic carbon reservoir are each available for each of these three cruises. Nutrient data is available for two of the three. Figure III.9 plots these parameters for deep samples (those for which salinity > 34.0) against distance from the head of the Saguenay Fjord. The corresponding linear regressions are all highly significant ($P > 99\%$) with the exception of the ^{13}C data, which is significant at 95%. Since these plots include most of the samples collected at depths > 200 m, the data include considerable variation that is due to depth. (This depth dependency is responsible for the scatter about the regression lines.)

Relative changes in the concentrations of oxygen and the nutrients over the length of the Laurentian Channel can then be calculated simply by comparing the slopes of the regression lines. It is possible to derive the amount of organic carbon oxidized from the corresponding change in the carbon isotope ratio of the dissolved inorganic carbon. Organic carbon is oxidized to CO_2 during the remineralization of organic matter. This results in a small increase in the concentration of dissolved inorganic carbon which is difficult to measure because of the high total concentration (≈ 2.2 mM). However, the organic matter oxidized is very different in carbon isotope ratio to the inorganic carbon. $\delta^{13}C$ values of organic carbon in the Gulf range from $- 22$ to $- 26°/_{oo}$, with

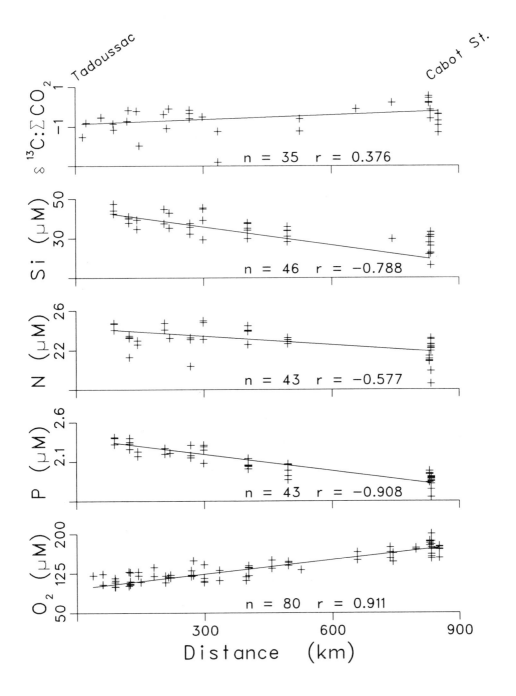

FIG. III.9. $\delta^{13}C$ (total CO_2), silicate, nitrate, phosphate, and dissolved oxygen in deep waters of the Laurentian Channel plotted against distance from the head of the Channel.

the particulate organic carbon found in deep Laurentian Channel water close to $-26^{0}/_{00}$ (see Chapter V); $\delta^{13}C$ for the inorganic carbon is $\approx 0^{0}/_{00}$. Knowing that the addition of carbon to the inorganic reservoir is small compared to the size of that reservoir and that the variation in the inorganic carbon isotope ratio is small compared to the difference in isotope ratio between organic and inorganic carbon, the gradient in carbon concentration can be calculated from the gradient in carbon isotope ratio by simple mass and isotope balance considerations:

$$dC/dx = C_{inorganic} \cdot (d\delta/dx) / (\delta_{organic} - \delta_{inorganic})$$

where C = carbon concentration
         δ = carbon isotope ratio
         x = distance along the Laurentian Channel

The relative changes produced by this analysis, combining the data from these three cruises, are $\Delta O : \Delta C : \Delta Si : \Delta N : \Delta P = 267 : 108 : 30 : 4.6 : 1$. Changes in oxygen, silicate, and phosphate are consistent with the values observed on the earlier cruises. The C:P ratio is very close to the open ocean value of 106:1 (Redfield et al. 1963). The N:P ratio is again very much lower than the open ocean value. (The small difference between the N:P ratio based on the earlier cruises and the estimate based on the more recent data is not significant due to the considerable scatter in the nitrate/distance plot.) As noted above, two possible explanations for the low N:P ratios may be incomplete regeneration of nitrate compared to phosphate or abnormally low nitrogen in the organic matter being oxidized. The consistency observed in the relative changes in oxygen and the nutrients from 1971 to 1975 shows that regeneration processes in the deep water of the Gulf of St. Lawrence are stable over extended time periods.

Dissolved oxygen measurements in the Gulf of St. Lawrence have been summarized by Dunbar et al. (1980). Two main features of the O_2 distributions come out of this summary. First, as already discussed, oxygen concentrations in the deep water decrease from Cabot Strait to the Estuary, reaching minimum values of 90 μM. Similar deep water trends are seen in the Esquiman Channel and in the channel to the north of Anticosti Island. An interesting exception to this trend to low dissolved oxygen concentration in the deep water is the observation of high deep oxygen values in the Mecatina area of the northwestern Esquiman Channel where a ridge isolates the deep water from the rest of the Esquiman Channel. This basin is apparently filled with water, with high O_2 content, entering the Gulf through the Strait of Belle Isle. Second, a summer oxygen maximum is found near the bottom of the thermocline and above the cold intermediate layer. This maximum results from equilibrium with the atmosphere at near freezing temperatures and is a residue of cold winter surface waters.

Nutrient Budgets

A nutrient budget for the Gulf of St. Lawrence calculated by Coote and Yeats (1979) is shown in Fig. III.10. In this budget, nutrient inputs from the St. Lawrence River and at Cabot Strait and the efflux at Cabot Strait were calculated for summer and winter conditions based on the three BIO nutrient surveys from 1971 to 1973. In addition, estimates were made of the amounts of nutrients transported vertically between the surface and deep layers. The main conclusion of this budget calculation is that with the phosphate budget in balance, there is a 0.9×10^6 t·yr^{-1} buildup of silicate and a 0.6×10^6 t·yr^{-1} loss of nitrate. It was concluded that the buildup of silicate is balanced by the removal of inorganic biogenic silica to Gulf sediments. The nitrate imbalance occurs predominantly in the surface layer and probably results from neglecting transports of nitrogen in the forms of dissolved organic nitrogen and ammonia.

Steven (1974) also made some estimates of fluxes in the Gulf. He estimated that 1.8×10^6 t of nitrate and 4×10^5 t of phosphate were transported into the Gulf at Pointe des Monts each year. These are 50% of the total inputs to surface waters of the whole Gulf estimated by Coote and Yeats (1979). Seaward transports in the surface layer at a section opposite Rimouski, Quebec, of 3×10^5 t of total nitrogen,

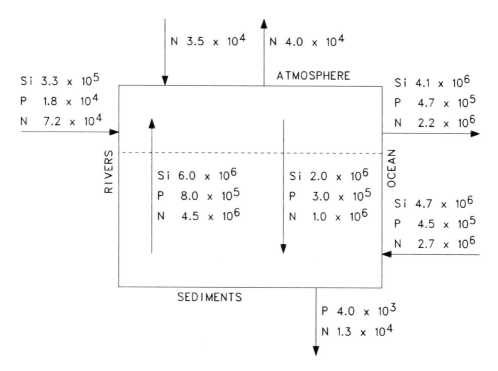

FIG. III.10. Nutrient budget for the Gulf of St. Lawrence (tonnes per year). (Redrawn from Coote and Yeats 1979).

4×10^4 t of total phosphorous, and 7×10^5 t of silicate for 5 months between May and September can be calculated from the data of Sinclair et al. (1976). These are considerably smaller than Stevens' estimate for the Pointe des Monts section. Greisman and Ingram's (1977) calculation of a surface nitrate flux through the Lower Estuary of 660 mole·s^{-1} corresponds to 2.4×10^4 t per month (June/July 1975) or about half the estimate based on Sinclair et al. mostly because Greisman and Ingram used a lower estimate of the freshwater discharge. These calculations show that the Estuary supplies a much smaller percentage of the overall nutrient input to Gulf surface waters than originally envisioned by Steven (1974). Since Steven had assumed that the Gaspe current extended across the entire width of the Estuary in making his estimate, this reduction is not surprising.

Dissolved silicate concentrations in the St. Lawrence River at Quebec City were measured on a monthly basis from May, 1974, to June, 1976, as part of a 2-year BIO study of the river chemistry. The monthly sampling gave a range of silicate concentrations between 12 and 83 μM. Maximum concentrations were found in April and May and minima from August to October. Monthly concentrations multiplied by monthly flow figures for the same time period give an average annual discharge of 5×10^5 t compared to the 3.3×10^5 t estimated by Coote and Yeats (1979).

It is evident from the data on nutrient distributions in the Upper St. Lawrence Estuary presented earlier in this chapter that the nutrient geochemistry in the Estuary will have an effect on the fluxes of nutrients through the Estuary. Although silicate mixes conservatively within the Estuary, significant decreases in silicate concentration were seen between Quebec City (location of most of the freshwater sampling) and the start of the salt wedge. If this behaviour, seen in both the 1974 cruises, is general it will result in a reduction of the silicate flux by about 20 %. The nitrate and phosphate fluxes on the other hand are augmented by 30 % by processes taking place in the Estuary. None of these estuarine processes will have a serious impact on the fluxes in the Gulf since the river flux is such a small component of the total flux through the Gulf.

References

BULLEID, E. R., AND D. M. STEVEN. 1972. Measurements of primary and secondary production in the Gulf of St. Lawrence. McGill Univ. Mar. Sci. Cent. MS Rep. 21: 111 p.

COOTE, A. R., AND R. S. HILTZ. 1975. Distribution of silicate, nitrate and phosphate in the Gulf of St. Lawrence. Bedford Inst. Oceanogr. Rep. BI-R-75-14: 65 p.

COOTE, A. R., AND P. A. YEATS. 1979. Distribution of nutrients in the Gulf of St. Lawrence. J. Fish. Res. Board Can. 36: 122-131.

DUNBAR, M. J., D. C. MACLELLAN, A. FILION, AND D. MOORE. 1980. The biogeographic structure of the Gulf of St. Lawrence. McGill Univ. Mar. Sci. Cent. Ms. Rep. 32: 142 p.

EL-SABH, M. I. 1975. Transport and currents in the Gulf of St. Lawrence. Bedford Inst. Oceanogr. Rep. BI-R-75-9: 180 p.

1979. The lower St. Lawrence estuary as a physical oceanographic system. Nat. Can. (Que.) 106: 55-73.

GREISMAN, P., AND G. INGRAM. 1977. Nutrient distribution in the St. Lawrence estuary. J. Fish. Res. Board Can. 34: 2117-2123.

INGRAM, R. G. 1979. Water mass modification in the St. Lawrence estuary. Nat. Can. (Que.) 106: 45-54.

KETCHUM, B. H., R. F. VACCARO, AND N. CORWIN. 1958. The annual cycle of phosphorus and nitrogen in New England coastal waters. J. Mar. Res. 17: 282-301.

LUCAS, J., AND P. CRITCH. 1974. Life in the oceans. Thames and Hudson, London, 216 p.

MORRIS, A. W., R.F.C. MANTOURA, A.J. BALE, AND R.J.M. HOWLAND. 1978. Very low salinity regions of estuaries: important sites for chemical and biological reactions. Nature (Lond.) 274: 678-680.

NEU, H. J. A. 1975. Runoff regulation for hydropower and its effect on the ocean environment. Can. J. Civil Eng 2: 583-591.

REDFIELD, A. C., B. H. KETCHUM, AND F. A. RICHARDS. 1963. The influence of organics on the composition of seawater, p 26-77. In M. N. Hill [ed.] The Sea, Vol 2, Wiley, New York, NY.

SEN GUPTA, R., AND F. KOROLEFF. 1973. A quantitative study of nutrient fractions and a stoichimetric model of the Baltic. Estuarine Coastal Mar. Sci. 1: 335-360.

SEVIGNY, J.-M., M. SINCLAIR, M.I. EL-SABH, S. POULET, AND A. COOTE. 1979. Summer plankton distributions associated with the physical and nutrient properties of the Northwestern Gulf of St. lawrence. J. Fish. Res. Board Can. 36: 187-203.

SINCLAIR, M., M. EL-SABH, AND J.-R. BRINDLE. 1976. Seaward nutrient transport in the lower St. Lawrence Estuary. J. Fish. Res. Board Can. 33: 1271-1277.

STEVEN, D. M. 1974. Primary and secondary production in the Gulf of St. Lawrence. McGill Univ. Mar. Sci. Cent. Ms Rep. 26: 116 p.

STEVEN, D. M., J. ACREMAN, F. AXELSON, M. BRENNAN, AND C. SPENCE. 1973. Measurements of primary and secondary production in the Gulf of St. Lawrence. Vol. II-IV. McGill Univ. Mar. Sci. Centre MS Rep. 23: 165 p., 24: 182 p., 25: 99 p.

SUBRAMANIAN, V., AND B. D'ANGLEJAN. 1976. Water chemistry of the St. Lawrence estuary. J. Hydrol. 29: 341-354.

SUTCLIFFE, W. H. JR. 1973. Correlations between seasonal river discharge and local landings of American lobster (Homarus americanus) and Atlantic halibut (Hippoglossus hippoglossus) in the Gulf of St. Lawrence. J. Fish. Res. Board Can. 30: 856-859.

THERRIAULT, J.-C., AND G. LACROIX. 1976. Nutrients, chlorophyll and internal tides in the St. Lawrence estuary. J. Fish. Res. Board Can. 33: 2747-2757.

YOUNG, W. 1976. Fluoride to chlorinity ratios in the waters of the St. Lawrence estuary and the Saguenay Fjord. Bedford Inst. Oceanogr. Rep. BI-R-76-1: 12 p.

CHAPTER IV

Organic Matter in the Gulf of St. Lawrence

Roger Pocklington

*Marine Chemistry Division, Physical and Chemical Sciences Branch,
Department of Fisheries and Oceans, Bedford Institute of Oceanography,
P.O. Box 1006, Dartmouth, N.S. B2Y 4A2*

Natural Organic Materials: Introductory Remarks

Knowledge of the organic chemical composition of the waters of the Gulf and its changes in space and time is currently limited to bulk measurements of total (TOC), particulate (POC), and dissolved organic carbon (DOC), supplemented by some measurements of particulate organic nitrogen (PON), plus a few organic contaminants. The stable isotope composition of organic carbon in the Gulf is discussed in Chapter V.

The quantity and chemical nature of the particulate organic matter (POM = POC multiplied by an empirical factor of 1.7 derived from CHN determination of mixed plankton; also similar factor with detritus) in any water body is an important consideration in calculation of the food sources available to non-photosynthesizing organisms, including fish. Phytoplankton production is the prime source of the POM (Parsons 1975), but vascular plants and attached algae, and water-borne and wind-borne materials of terrestrial origin may make a significant contribution to the POM of marginal seas such as the Gulf. Not all the organic matter synthesized by phytoplankton is in fact in particulate form; extracellular release of dissolved organic matter (DOM) by healthy organisms during normal growth, and upon their death is proven. (The direct contribution by algae to the dissolved organic pool is on the order of 25 % of primary production as against 75 % as particulate organic matter, according to Wafer et al. 1984).

The sampling methods and analytical procedures used for POM and DOM studies in the Gulf have been described by MacKinnon (1978), Gershey et al. (1979), and Pocklington and Kempe (1983). Thousands of water samples collected at a wide variety of locations within the Gulf have been analyzed over the last two decades. Particles were collected on pre-combusted membranes of defined pore size (0.8 μm silver membranes; Selas Flotronics, Silver Springs, Maryland). In studies of the dissolved fraction, total organic carbon (TOC) is the quantity actually measured by the method of MacKinnon (1978), but if POC is measured independently by the method of Pocklington and Kempe (1983), the true dissolved concentration may be calculated by difference (DOC = TOC - POC). Sediment samples taken by surface grab were analyzed by a similar procedure for SOM (sedimentary organic matter; Pocklington 1973; Pocklington and Hagell 1975; Pocklington and Morash 1979).

General Features of Organic Matter Distribution Within the Gulf and Adjacent Ocean Waters

Concentrations of POC and PON within the Gulf are shown in Table IV.1.

On the basis of physical criteria, the waters of the Gulf are divisible into three layers, two of which are always present: from the surface to 40-100 m (surface layer) and from 150 m to the bottom (deep layer). In most months (see Chapter I), these two persistent layers are separated by an intermediate layer within which temperature does not exceed 1.5°C (cold layer). The deep layer maintains a constant temperature throughout the year; the surface layer warms seasonally leaving the cold layer — the remnants of the previous winter's cooling — sandwiched between them. The waters of this intermediate cold layer mix with the surface layer early in the following year. Enrichment of the surface layer by vertical mixing, entrainment, and/or upwelling provides the supply of nutrients for phytoplankton production, initiating and sustaining levels of annual

TABLE IV.1. Representative monthly data in each layer.

Month	Depth range (m)	No. of samples	POC mean (range)	PON mean (range)	C/N mean (range) atom
			mg·m^{-3}		
Jan.	0-100	69	20 (7- 46)	2.8 (1.3- 6.0)	8.3 (5.0-14.0)
	75-170	22	12 (7- 22)	1.6 (0.9- 3.2)	9.2 (7.1-17.9)
	145-450	33	12 (7- 34)	1.4 (0.6- 3.4)	9.9 (5.7-18.8)
Apr.	0- 50	93	147 (24-256)	19.7 (6.6-79.0)	8.7 (6.0-13.1)
	20-200	141	31 (13-533)	3.8 (1.2-62.4)	9.7 (6.1-14.8)
	200-400	65	18 (12- 39)	1.9 (1.2- 3.2)	11.5 (8.8-16.7)
June	0- 50	85	65 (23-256)	9.5 (3.2-45.0)	7.6 (6.3-11.6)
	40-200	84	28 (12-277)	3.6 (1.5-26.9)	9.1 (6.7-21.0)
	150-450	29	22 (10- 38)	2.4 (1.3- 4.2)	10.5 (6.9-21.4)
Aug.	0- 40	97	99 (34-412)	13.3 (5.0-58.6)	8.7 (6.1-15.0)
	20-200	180	21 (13- 51)	2.7 (1.8- 7.9)	8.9 (6.8-10.8)
	200-450	50	20 (13- 33)	2.1 (1.1- 4.5)	11.4 (7.6-19.1)
Nov.	0- 75	22	47 (30- 97)	6.3 (3.9- 8.4)	8.5 (7.0-13.0)
	100-250	12	23 (13- 44)	2.4 (1.2- 5.2)	11.0 (9.1-13.5)

primary production within certain regions of the Gulf (e.g. in the Gaspé Current, 385 g C·m^{-2}·yr^{-1}, Steven 1975) comparable with those in the most productive regions outside it (e.g. George's Bank, 375 g C·m^{-2}·yr^{-1} — Sambrotto et al. 1984).

Sustained levels of primary organic production are maintained within much of the Gulf by subsurface water of high nutrient content flowing in at Cabot Strait, moving along the axis of the deep channels and mixing into the surface layer by upwelling (Bugden 1981) and/or entrainment (Neu 1982a,b). These processes are particularly marked at the head of the Laurentian Channel where the bottom shoals sharply from 350 to 40 m in only 20 km; they are not so apparent over the Magdalen Shallows or in the northern Gulf.

Organic matter concentration and composition show systematic variations in the course of a year (Table IV.1). In the surface layer, concentrations of POC and PON peak in the spring (April-May), decline in early summer (June-July), and increase again in late summer (August), though not to levels as high as those in the spring. Composition of the POM is indicated by the C/N ratio which is lowest in early summer as a result of fresh organic production at this season, but otherwise relatively invariant.

Concentrations of POC and PON decline in the fall (November) and are lowest in winter (January), when they are no higher than in waters external to the Gulf (Pocklington 1985a). In contrast, the concentrations of POC and PON in the Gulf in spring and summer (April-August) are as high as in one of the most productive areas of the world ocean — off the northwest African coast during the months of intense upwelling (Pocklington and MacKinnon 1982).

In the intermediate layer, there is an increase in POC and PON early in the year (between January and April) but no further quantitative change until a decrease in late summer. A decrease in the mean C/N ratio of the POM (between April and August) is a consequence of freshly produced POM replacing that of a previous annual cycle ("older" POM has a higher C/N ratio than "younger"). In the deep layer, there is, other than lower concentrations of POC and PON during the winter, little change in either concentration or composition of POM over the year. As much of the deep water originates outside the Gulf, it would not be expected to reflect changes within the system.

The mass balance of organic matter (OM) within the system has been computed since water fluxes in certain key sections of the Gulf (Fig. IV.1) are known or have been calculated from available data. At Cabot Strait, surface salinity reaches maximum values

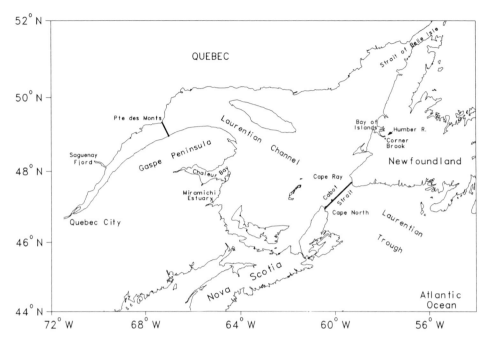

FIG. IV.1. Gulf of St. Lawrence showing sections and other locations mentioned in the text.

for the year in January, due in part to the release of salt when ice is formed and in part to a decrease in freshwater runoff in the winter. Water movements in Cabot Strait are dominated by an outward surface flow, strongest around Cape North on the Nova Scotian side, and an inward flow off Cape Ray, Newfoundland side. Through this section, a volume of water of low salinity, some 20-30 times the total annual freshwater input and equivalent to the volume of water on the Nova Scotian Shelf between the Laurentian Channel and George's Bank, leaves the Gulf over a year. To compensate, water of higher salinity from the North Atlantic enters at Cabot Strait and flows along the axes of the deep channels.

A simple box diffusion model of the carbon cycle within the Gulf of St. Lawrence has been constructed and is shown in Fig. IV.2. Units of millions of metric tonnes per year are used in the figure and throughout this discussion. The model reproduces the dissolved and particulate organic carbon fluxes between the Gulf and other reservoirs on an annual time scale, which necessarily limits the complexity of the model. These fluxes may not in fact be balanced annually, but obviously are on a longer time scale.

Fresh Water Inputs

The largest single source of freshwater to the Gulf is the St. Lawrence River. Data for TOC and POM collected bimonthly at Quebec City over the years 1981-85 permit the calculation of the fluvial contribution of OM to the Gulf (Pocklington and Tan 1987). A number of attempts have been made to estimate total freshwater inputs to the Gulf; we accept Bugden's (1981) figure of 500 $km^3 \cdot yr^{-1}$. Annual precipitation at these latitudes is balanced by evaporation within the standard deviation of annual means.

The annual discharge of the St. Lawrence River (413 km^3) contains organic matter in particulate (POC; 236-367 $\times$ 10^3 t; range 1981-1985) and dissolved (DOC; 1.29-1.72 $\times$ 10^3 kt) form. The concentration of POC (and particulate organic nitrogen) is linearly correlated with discharge (increased during the spring flood and the fall enhancement of flow), but concentration of DOC is not so simply related to discharge. In consequence, the total organic carbon (1.66-2.00 $\times$ 10^3 kt) discharged

51

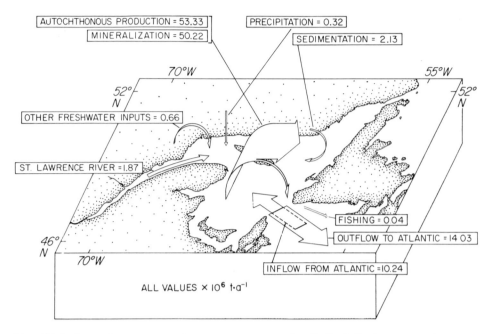

FIG. IV.2. Annual transports of organic matter in the Gulf of St. Lawrence (in units of 10^6t C·yr^{-1}). (Updated from Pocklington 1986, with permission).

annually by the river is relatively invariant. It is a substantial quantity of organic matter on a world scale — more than three times the annual discharge of TOC from the Columbia River to the Pacific Ocean (0.59×10^3 kt; Dahm et al. 1981) and more than half that of the Mississippi River to the Gulf of Mexico (3.4×10^3 kt; Dahm et al. 1981). Even so, it is less than 30 % of the annual photosynthetic production in the St. Lawrence Estuary, the most productive region in the Gulf of St. Lawrence (6.6×10^3 kt C; Steven 1975), into which the river flows. The gross outflow of organic matter from the Estuary is nearly six times the river influx and equivalent to the estimated annual production of organic carbon in the Gaspé Current (11×10^3 kt; Steven 1975). The net export of TOC (Table IV.2) is of the same order as the river input. Pocklington and Tan (1987) conclude that the organic matter contributed annually by the St. Lawrence River is more than adequate to account for the organic matter sedimenting within the Estuary (see Chapter V).

TABLE IV.2. Mass transports through the Estuary section in different months.

	Out				In			
Month	Water[a] (km³)	DOC (10³ t)	POC (10³ t)	PON (10³ t)	Water[a] (km³)	DOC (10³ t)	POC (10³ t)	PON (10³ t)
Jan	906	1075	17.4	2.36	839	793	9.8	1.40
July	710	800	36.1	5.53	667	638	17.2	2.47
Nov	1189	1296	42.6	5.61	1153	1279	39.1	5.03

[a] Volume transports for July and November from El-Sabh (1975).

Exchanges with the Adjacent Ocean

The greatest proportion of OM leaving the Gulf is in the water which flows out in the surface layer around Cape North. An estimate of the quantity of OM contributed to the Atlantic Ocean from the Gulf through Cabot Strait can be made by combining OM concentrations with geostrophic volume transports (Table IV.3) adjusted to satisfy the condition of zero net mean salt transport through the section (El-Sabh 1977). Although El-Sabh's calculations are the most appropriate data available for the construction of a budget of this type, they assumed that fluxes through the Strait of Belle Isle were insignificant. Very recent work seriously questions this assumption (see Chapters I and X).

The gross annual seaward transport of TOC is one quarter of the autochthonous production in the Gulf as calculated by Steven (1975) and three times as much as is contributed by organic carbon fixation on the Nova Scotian Shelf to the adjacent North Atlantic (5.25×10^3 kt·yr^{-1}) calculated from data of Fournier et al. (1977). It is of the same order as the very uncertain export of OM by the Amazon River (18×10^3 kt C·yr^{-1}; Richey et al. 1980). The nature of the organic matter imported is not the same as that which is exported: the mean C:N ratio of organic matter in inflowing water is significantly higher than that in outflowing water and implies a net loss of 0.5×10^3 kt N·yr^{-1} from the Gulf in organic form. This is very close to an imbalance of 0.6×10^3 kt N·yr^{-1} which Coote and Yeats (1979) computed from their nutrient budget of the Gulf (see Chapter III).

TABLE IV.3. Mass transports through Cabot Strait.

Months	Water[a] (10^3 km^3)	DOC (10^6 t)	POC (10^3 t)	PON (10^3 t)	C:N (atoms)
			Out		
Feb. - Mar.[b]	1.70	1.77	60.0	7.25	9.65
Apr. - May[b]	1.61	1.83	158.5	20.49	9.02
June - July	1.86	2.13	139.4	19.69	8.26
Aug. - Sept.[b]	2.82	3.26	145.0	19.45	8.69
Oct.[b] - Nov.	2.37	2.54	107.4	15.70	7.97
Dec.[b] - Jan.	2.12	1.85	53.0	6.75	9.15
Annual Total	12.48	13.37	663.2	89.34	8.66
			In		
Feb. - Mar.[b]	1.69	1.42	25.0	3.10	9.40
Apr. - May[b]	1.48	1.32	60.6	7.34	9.63
June - July	1.78	1.42	66.4	7.51	10.32
Aug. - Sept.[b]	2.73	2.30	94.4	8.32	13.24
Oct.[b] - Nov.	2.23	1.89	75.7	9.42	9.37
Dec.[b] - Jan.	2.08	1.52	35.7	4.57	9.10
Annual Total	11.99	9.88	357.8	40.26	10.36

[a] From El-Sabh (1977) as modified by Bugden (1981).
[b] Interpolated month.

Exchanges with Sediments

Sediments in the Gulf are generally low in OM as compared with sediments beneath other areas of high organic productivity (Loring and Nota 1973; Pocklington 1986).

TABLE IV.4. Organic carbon, organic nitrogen, and C:N ratios in the major sediment types in different areas of the Gulf of St. Lawrence. (Data from Pocklington and Morash 1979).

	Organic C		Organic N		C:N	
	Mean	Range	Mean	Range	Mean	Range
		$(mg \cdot g^{-1})$		$(mg \cdot g^{-1})$		(atoms)
Upper Estuary						
True sands	1.0	0.6- 1.6	0.13	0.11-0.26	8.9	5.4-15.5
Pelitic sands	7.4	5.8- 8.5	0.73	0.58-1.00	11.9	9.1-15.6
Pelites	18.2	10.3-29.4	1.56	1.00-2.66	13.6	11.9-15.9
Saguenay Fjord						
Sands	4.2	0.5- 8.0	0.32	0.00-0.63	15.3	7.5-21.2
Pelites	25.7	11.7-37.8	1.42	0.61-1.85	21.1	15.3-42.8
Lower Estuary						
Sands	4.1	1.9- 6.3	0.35	0.15-0.83	13.8	12.0-15.9
Pelites	14.8	11.8-19.3	1.36	0.92-1.61	12.7	9.5-16.4
Open Gulf [a]						
Sands	7.3	4.5- 8.6	0.76	0.36-1.03	11.2	7.6-14.6
Pelites	13.6	9.9-20.5	1.72	0.76-3.28	9.2	6.1-15.2
Outside Gulf [b]						
Sands	4.6	1.8- 8.0	0.59	0.18-1.29	9.1	6.0-11.7
Pelites	21.9	12.3-48.1	3.00	1.33-4.54	8.5	6.9-12.4

[a] North of 48°N, west of 62°W, excluding Chaleur Trough and Jacques Cartier Passage.
[b] Laurentian Channel outside Cabot Strait.

This may be due to the large area of high-energy environments in which only coarse-grained sediments can settle. Concentration of organic matter is highly dependent on the grain size of the sediments. Table IV.4 lists POC and PON concentrations and C:N atom ratios for different grain sizes and different regions of the Gulf. POC concentrations in sands range from 0.6 to 8.6 $mg \cdot g^{-1}$ in the region. In pelites (i.e. sediments of grain size < 50 μm), POC varies over the range 9.9 to 37.8 $mg \cdot g^{-1}$. Much higher concentrations can be found locally inshore near organic matter sources. For example, Pocklington (1975) reported values of 180 $mg \cdot g^{-1}$ in Corner Brook harbor; Rashid and Reinson (1979) reported POC concentrations as high as 67 $mg \cdot g^{-1}$ in the Miramichi Estuary. The contribution of TOC to the sediments is 2.13×10^3 kt$\cdot$yr^{-1} (Fig. IV.2).

The fate of most (76 % by difference — see Fig. IV.2) of the OM within the Gulf is not to be exported or buried in the sediments, but to be mineralized and, as inorganic nutrients, returned to the surface layer. This process consumes oxygen in the deep water, reducing the O_2 concentration by one half between Cabot Strait and the Estuary (6.72×10^{12} moles O_2 over the whole Gulf; Coote and Yeats 1979; and Chapter III). Were it not for a vigorous vertical circulation (Bugden 1981), deep water within the Gulf would become anoxic, as in the Baltic, within a year (Pocklington 1986). This then is the natural system upon which anthropogenic organic materials impinge.

Anthropogenic Organic Materials

Quantitatively, human additions of organic compounds to the Gulf are insignificant, amounting to less than 0.005 % of annual primary production (Fig. IV.2). Qualitatively, they may give some cause for concern. A 10-year monitoring program for oil (Chapter IX) has shown a decline in the background levels of dissolved and dispersed petroleum residues over the decade. This is due, perhaps, to pollution abatement measures which were instituted during the early 1970's. A steady state has been reached;

this suggests that the capacity of the Gulf to assimilate these anthropogenic substances is now in balance with present rates of input. However, there is no room for complacency because an increase in the rate of input could easily upset this equilibrium with ensuing deterioration of environmental quality.

Chlorinated hydrocarbons, DDT-group pesticides, and PCBs, being exclusively man-made, can be directly related to human activity. In the waters of the Gulf, determinations of concentrations of chlorinated hydrocarbons have been made by Smillie (1976). He measured surface water concentrations at 14 stations from the Gulf and Lower St. Lawrence Estuary. The precision of these measurements was limited by the packed column chromatography available at the time and by the low levels observed (DDE and DDT concentrations were always below the method's detection limit, estimated at 0.2 ng$\cdot$L^{-1}), but the results do show that PCB concentrations (< 2.1 ng$\cdot$L^{-1}) are very low in comparison to concentrations in other areas (e.g. μg$\cdot$L^{-1} in the Baltic, Ehrhardt 1981). Since toxic effects of chlorinated hydrocarbons are not observed until concentrations reach the μg$\cdot$L^{-1} level in seawater (Ehrhardt 1981), present concentrations of PCBs and DDT group compounds do not appear to be an acute problem in the Gulf in general though they might be in certain locations (Couillard 1982) such as Lake Saint-Louis and Lake Saint-François which drain via the St. Lawrence River to the Gulf.

A few measurements of DDT pesticides and PCB's have also been made in surface sediments of the Gulf. Leonard (1977) reported PCB concentrations up to 12 ng$\cdot$g^{-1} in samples from the Northumberland Strait area and the Miramichi Estuary. Total DDT group concentrations were < 2 ng$\cdot$g^{-1} in samples from the Northumberland Strait but ranged up to 38 ng$\cdot$g^{-1} in the Miramichi. A number of analyses performed in advance of dredging operations have also been reported for sediments in harbors from the Gulf (Travers and Wilson 1977).

Some limited information is also available on polycyclic aromatic hydrocarbons (PAH's). Martel et al. (1986) measured PAH's in sediment samples from the Saguenay Fjord. They found a strong gradient in total PAH concentrations from 4.5 μg$\cdot$g^{-1} in the upper reaches of the Saguenay to 0.5 μg$\cdot$g^{-1} in the eastern basin. They attribute these high concentrations to waste from the aluminum smelting industry along the Saguenay River, and note that there may be decreasing levels in the most recent surface sediments (see also Chapter VIII). Measurements on PAH's in mussels in the Gulf were reported by Cossa et al. (1983) and Picard-Berube et al. (1983).

Waste from pulp and paper mills, of which there are 50-100 in the drainage basin of the Gulf of St. Lawrence, has been shown to be a good general indicator of human effects upon enclosed waters. Lignin, an unequivocal indicator of terrigenous organic matter (Pocklington and Roy 1975), is an aromatic polymer of phenyl propane units yielding vanillin in consistent yield upon nitrobenzene oxidation (Pocklington and Mac-Gregor 1973). Lignin contributes 28 % of the weight of dry spruce wood and is essentially undegraded in anoxic sediments. Pocklington (1976) has discussed lignin analyses from the Gulf; Rashid and Reinson (1979) have discussed the importance of pulp mill wastes to the organic matter in nearshore sediments in the Miramichi. Table IV.5 lists the lignin concentrations found in some representative environments in the Gulf. Concentrations are highest in areas receiving relatively direct inputs of pulp mill waste such as the upper Saguenay Fjord and Corner Brook harbor. Note that the lignin concentration in the Humber River is much lower than that in Corner Brook harbor into which it flows. This observation suggests that high lignin concentrations are indicative of highly concentrated organic waste, rather than solely an indicator of terrigenous organic matter. Lower concentrations of lignin are found in areas somewhat more distant from these inputs of organic wastes such as the Chaleur Trough and the St. Lawrence Estuary. Lignin concentrations in the open areas of the Gulf are close to zero.

In surficial sediments of the Miramichi Estuary, Rashid and Reinson (1979) found a much higher quantity of organic matter than would be expected to occur in an estuary which is shallow and well-mixed. Enhanced organic carbon concentration followed the distribution of fine-grained sediment, although higher concentrations occurred in sedi-

TABLE IV.5. Contribution of lignin to sediments from the Gulf of St. Lawrence (adapted from Pocklington 1976).

Location	Lignin (mg·g⁻¹)	Organic Matter[a] (%)	Lignin (as % of O.M.)
Upper Saguenay	4.56	5.54	8.2
	9.04	7.13	12.7
Lower Saguenay	0.03	0.64	0.5
St. Lawrence Estuary	0.77	3.61	2.1
Laurentian Channel (inside Gulf)	0	2.60	0
Chaleur Trough	0.71	4.04	1.8
Esquiman Channel	+[b]	4.83	0
Corner Brook Harbour	16.51	17.2	9.6
Humber River	1.46	4.19	3.5
Cabot Strait	0	3.62	0
Laurentian Trough (outside Gulf)	0	3.26	0

[a] Organic carbon converted to organic matter using the empirical multiplier 1.887.

[b] + indicates positive spot test but no quantitative measure.

ments of the drowned river channel than in sediments of the bay portion of the Estuary. The high quantity of organic matter in all the sediments of the Estuary, and differences in concentrations between river channel and bay sediments, are the direct result of the discharge of pulp-mill effluent into the upper reaches of the Estuary. The organic carbon isotopic composition of the sediments suggests that land-derived material is the predominant source of organic matter throughout the estuary. Marine organic material is restricted to the sediments near the estuary mouth. Dispersal of the organic matter is achieved largely by transport in the upper freshwater effluent layer, and by settling through the water column when transport energies are reduced. Settling of organic matter was highest in areas where deposition of fine-grained inorganic sediments prevails. In the Saguenay Fjord, inputs of lignified material from pulp and paper plants have been traced not only longitudinally along the axis of the Fjord, but down the sedimentary column where they persist for at least one hundred years (see Chapter VIII).

References

BUGDEN, G. L. 1981. Salt and heat budgets for the Gulf of St. Lawrence. Can. J. Fish. Aquat. Sci. 38: 1153–1167.

COOTE, A. R., AND P. A. YEATS. 1979. Distribution of nutrients in the Gulf of St. Lawrence. J. Fish. Res. Board Can. 36: 122–131.

COSSA, D., M. PICARD-BERUBE, AND J.P GOUYGOU. 1983. Polynuclear aromatic hydrocarbons in mussels from the Estuary and northwestern Gulf of St. Lawrence, Canada. Bull. Environ. Contam. Toxicol. 31: 41–47.

COUILLARD, D. 1982. Evaluation des teneurs en composés organochlores dans le fleuve, l'estuaire et le golfe Saint-Laurent, Canada. Environ. Pollut. (Series B) 3: 239–270.

DAHM, C. N., S. V. GREGORY, AND P. K. PARK. 1981. Organic carbon transport in the Columbia River. Estuarine Coastal Shelf Sci. 13: 645–658.

EHRHARDT, M. 1981. Organic substances in the Baltic Sea. Mar. Pollut. Bull. 12: 210–213.

EL-SABH, M. I. 1975. Transport and currents in the Gulf of St. Lawrence. Bedford Inst. Oceanogr. Rep. BI-R-75-9: 180 p.

1977. Oceanographic features, currents, and transport in Cabot Strait. J. Fish. Res. Board Can. 34: 516–528.

FOURNIER, R. O., J. MARRA, R. BOHRER, AND M. VANDET. 1977. Plankton dynamics and nutrient enrichment of the Scotian Shelf. J. Fish. Res. Board Can. 34: 1004-1018.

GERSHEY, R. M., M. D. MACKINNON, P. J. LEB. WILLIAMS, AND R. M. MOORE. 1979. Comparison of three oxidation methods used for the analysis of the dissolved organic carbon in seawater. Mar. Chem. 7: 289-306.

LEONARD, J. D. 1977. Organohalogens in coastal sediments from the Maritime provinces, Canada. Bedford Inst. Oceanogr. Rep. BI-R-77-6: 12 p.

LORING, D. H., AND D. J. G. NOTA. 1973. Morphology and sediments of the Gulf of St. Lawrence. Bull. Fish. Res. Board Can. 182: 147 p.

MACKINNON, M. D. 1978. A dry oxidation method for the analysis of the TOC in seawater. Mar. Chem. 7: 17-37.

MARTEL, L., M. J. GAGNON, R. MASSE, A. LECLERC, AND L. TREMBLAY. 1986. Polycyclic aromatic hydrocarbons in sediments from the Saguenay Fjord, Canada. Bull. Environ. Contam. Toxicol. 37: 133-140.

NEU, H. J. A. 1982a. Man-made storage of water resources — a liability to the ocean environment? Part I. Mar. Pollut. Bull. 13: 7-12.

        1982b. Man-made storage of water resources — a liability to the ocean environment? Part II. Mar. Pollut. Bull. 13: 44-47.

PARSONS, T. R. 1975. Particulate organic carbon in the sea, p. 365-383. In J. P. Riley and G. Skirrow [ed.] Chemical Oceanography. Vol. 2, 2nd edition. Academic Press, London.

PICARD-BERUBE, M., D. COSSA, AND J. PIUZE. 1983. Teneurs en benzo 3,4 pyrene chez Mytilus edulis L. de l'Estuaire et du Golfe du Saint-Laurent. Mar. Environ. Res. 10: 63-71.

POCKLINGTON, R. 1973. Organic carbon and nitrogen in sediments and particulate matter from the Gulf of St. Lawrence. Bedford Inst. Oceanogr. Rep. BI-R-73-8: 16 p.

        1975. Carbon, hydrogen, nitrogen and lignin determinations on sediments from the Gulf of St. Lawrence and adjacent waters. Bedford Inst. Oceanogr. Rep. BI-R-75-6: 12 p.

        1976. Terrigenous organic matter in surface sediments from the Gulf of St. Lawrence. J. Fish. Res. Board Can. 33: 93-97.

        1985a. Organic matter in the Gulf of St. Lawrence in winter. Can. J. Fish. Aquat. Sci. 42: 1556-1561.

        1986. The Gulf of St. Lawrence and the Baltic Sea: two different organic systems. Dt. hydrogr. Z. 39: 65-75.

POCKLINGTON, R., AND G. T. HAGELL. 1975. The quantitative determination of organic carbon, hydrogen, nitrogen and lignin in marine sediments. Bedford Inst. Oceanogr. Rep. BI-R-75-18: 16 p.

POCKLINGTON, R., AND S. KEMPE. 1983. A comparison of methods for POC determination in the St. Lawrence River. Mitt. Geol.- Palaont. Inst. Univ. Hamburg 55: 145-151.

POCKLINGTON, R., AND C. D. MACGREGOR. 1973. The determination of lignin in marine sediments and particulate form in seawater. Intern. J. Environ. Anal. Chem. 3: 81-93.

POCKLINGTON, R., AND M. D. MACKINNON. 1982. Organic matter in upwelling off Senegal and The Gambia. Rapp. P.-V. Reun. Cons. Int. Explor. Mer 180: 254-265.

POCKLINGTON, R., AND L. MORASH. 1979. Organic carbon, nitrogen and lignin in sediments from the Gulf of St. Lawrence and adjacent waters. Bedford Inst. Oceanogr. Rep. BI-R-79-1: 14 p.

POCKLINGTON, R., AND S. ROY. 1975. Utilisations de composés organiques comme traceurs de materiel d'origine terrestre dans l'océan. Int. Coun. Explor. Sea Hydrography Committee CM 1975/c:10.

POCKLINGTON, R., AND F. C. TAN. 1987. Seasonal and annual variations in the organic matter contributed by the St. Lawrence River to the Gulf of St. Lawrence. Geochim. Cosmochim. Acta 51: 2579-2586.

RASHID, M. A., AND G. E. REINSON. 1979. Organic matter in surficial sediments of the Miramichi Estuary, New Brunswick, Canada. Estuarine Coastal Mar. Sci. 8: 23-36.

RICHEY, J. E., J. T. BROCK, R. J. NAIMAN, R. C. WISSMAR, AND R. F. STALLARD. 1980. Organic carbon oxidation and transport in the Amazon River. Science (Wash., D.C.) 207: 1348-1351.

SAMBROTTO, R. M., J. J. GOERING, AND C. P. McROY. 1984. Large yearly production of phytoplankton in the western Bering Strait. Science (Wash., D.C.) 225: 1147-1150.

SMILLIE, R. D. 1976. Trace concentrations of organochlorines from waters of the Gulf of St. Lawrence. Bedford Inst. Oceanogr. Rep. BI-R-76-18: 6 p.

STEVEN, D. M. 1975. Biological production in the Gulf of St. Lawrence, p. 229-248. In T. W. M. Cameron and L. W. Billingsley [ed.] Energy flow — Its biological dimension, Royal Society of Canada, Ottawa, 319 p.

TRAVERS, I. C., AND R. C. H. WILSON. 1977. PCB's in the Atlantic provinces. Environmental Protection Service Surveillance Report EPS-5-AR-77-13: 75 p.

WAFAR, M., P. LE CORRE, AND J.-L. BIRRIEN. 1984. Seasonal changes of dissolved organic matter (C, N, P) in permanently well-mixed temperate waters. Limnol. Oceanogr. 29: 1127-1132.

CHAPTER V

Stable Isotope Studies in the Gulf of St. Lawrence

F. C. Tan and P. M. Strain

*Marine Chemistry Division, Physical and Chemical Sciences Branch,
Department of Fisheries and Oceans, Bedford Institute of Oceanography,
P.O. Box 1006, Dartmouth, N.S. B2Y 4A2*

Terminology

A brief description of the terms and scale conventions used in geochemical applications of stable isotope ratio measurements follows for those not familiar with the field. Natural variations in the abundance of the rare stable isotopes are reported as δ-values on a scale defined by the equation:

$$\delta\,(^o/_{oo}) \;=\; 1000\;\left(\frac{R_{sample}}{R_{standard}} - 1\right)$$

where R is the ratio of the less common (and heavier) isotope to the most common isotope — i.e. $^{13}C/^{12}C$ and $^{18}O/^{16}O$ for the oxygen and carbon studies discussed below. Analytical considerations dictate that the isotope ratio of a standard is included in the definition of this scale. The standards used are PDB (a Belemnite from the Pee Dee formation) for carbon and SMOW (standard mean ocean water) for oxygen isotopes in water. The corresponding measurements are referred to as $\delta^{13}C_{PDB}$ and $\delta^{18}O_{SMOW}$ values (the standards may not be explicitly included). The isotope ratios of both the carbon and oxygen standards fall near the upper end of the range of values discussed below. This means that most, but not all, of the isotope ratios discussed are negative, a fact that leads to considerable confusion in discussing trends. One isotope ratio may be said to be higher or heavier than another: both terms mean that the first δ-value is more positive than the second. The term heavier, which may be applied to the carbon (oxygen) in a sample as well as to the isotope ratio, arises because the average atomic weight for the element in question is higher in samples containing more of the heavy isotope.

Organic Carbon Isotopes

Introduction

The isotopic composition of organic carbon in the Gulf of St. Lawrence has been described in a number of studies. This work includes the use of the isotope ratio of sedimentary organic carbon to map the distribution of terrestrial organic matter in Gulf of St. Lawrence sediments (Tan and Strain 1979b), a similar study on the surface sediments in the Miramichi Estuary (Rashid and Reinson 1979), a report on the carbon isotope composition of particulate organic carbon (Tan and Strain 1979a), a detailed study into the sources and sinks of organic carbon in the St. Lawrence Estuary (Tan and Strain 1983), and three reports on the isotopic composition of organic carbon in the St. Lawrence River discharge (Pocklington and Tan 1983; Tan 1988; Pocklington and Tan 1987). Rather than discuss these studies chronologically, they are reviewed below on a regional basis.

Upper St. Lawrence Estuary

The distribution of terrestrial organic matter in estuarine and marine sediments has been the subject of numerous investigations worldwide. Techniques used to trace terrestrial organic matter in the marine environment include C/N, $^{13}C/^{12}C$, deuterium/hydrogen, and $^{15}N/^{14}N$ ratios and lignin concentrations (Sackett 1964; Nissenbaum and Kaplan 1972; Gardner and Menzel 1974; Nissenbaum 1974; Pocklington 1976; Hedges and Parker 1976). The $^{13}C/^{12}C$ method is based upon a general enrichment of ^{12}C in terrestrial organic matter compared with marine organic matter (Craig 1953; Smith and Epstein 1971; Eckelman et al. 1962; Sackett 1964; Parker et al. 1972; Newman et al. 1973).

Figure V.1 shows the distribution of the carbon isotope ratios of the organic carbon in surface sediments of the St. Lawrence Estuary as reported in Tan and Strain (1979b). Organic carbon concentrations were also measured in this study, with most values between 1 and 2% (dry weight). Organic carbon levels are closely related to grain size, with higher concentrations associated with finer sediments (see Chapter VII). More extensive data for sedimentary organic carbon concentrations are given by Pocklington (1973, 1975).

The small range of $\delta^{13}C$ values observed in the Upper Estuary is a result of the trapping of suspended particulate matter within the turbidity zone (see Chapter II). The mean carbon isotope ratio for the six Upper Estuary samples ($-25.0 \pm 0.6\%$) was not significantly different from the average of three samples ($-25.4 \pm 0.2\%$) from the St. Lawrence River collected 0–70 km upstream of Quebec City.

On the basis of both the narrow ranges and the similarity between the mean isotope ratios from these two areas, Tan and Strain (1979b) concluded that the organic carbon deposited within the Upper Estuary is almost entirely of terrestrial origin. Pocklington and Leonard (1979) challenged this conclusion on the basis of C/N ratio data for 25 samples. They calculated that the amount of terrestrial material found in Upper Estuary sediments ranged from 3 to 50%. Both these studies suffered from inadequate characterization of the organic matter delivered by the St. Lawrence River to the Estuary, and from the failure to consider that the riverborne material might be a mixture of terrestrial carbon and carbon produced *in situ* in the river.

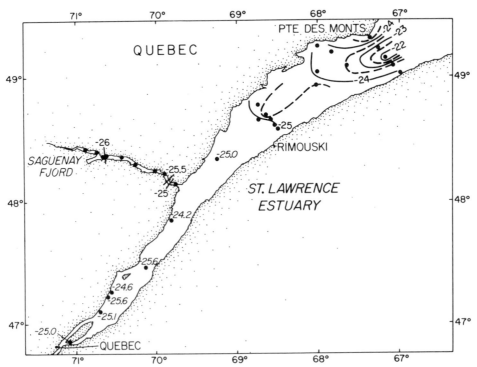

FIG. V.1. $\delta^{13}C$ distribution for organic carbon in surface sediments of the St. Lawrence Estuary and the Saguenay Fjord (from Tan and Strain 1979b with permission).

The resolution of this controversy became one of the aims of a long-term monitoring program on the St. Lawrence discharge (Pocklington and Tan 1983; Pocklington 1985b; Tan 1988; Pocklington and Tan 1987). Twice monthly samples of riverborne POC were analyzed over a 5-year period for both carbon isotope and C/N ratios (Fig. V.2). This detailed characterization of the riverborne carbon showed that it is generally dominated by terrestrial material except in the summer (July to September) when an important contribution from primary production in the river produces heavier isotope ratios and lower C/N ratios. Both the isotope and C/N ratios of the riverborne material are consistent with what had previously been described in Upper Estuary sediments. Pocklington and Tan (1987) used the accompanying carbon concentration data to calculate that the sedimentation of only one third of the river-derived organic carbon (300 kt·a^{-1}) would be sufficient to account for all of the organic sedimentation in the Upper Estuary (91 kt·a^{-1}; see also Chapter IV). They concluded that terrigenous organic material dominates the fraction buried in fine sediments of the Upper St. Lawrence Estuary.

In a study investigating the behaviour of carbon in the St. Lawrence Estuary, Tan and Strain (1983) measured carbon isotope ratios of total particulate organic carbon (0.8 μm silver filters), planktonic organic carbon (vertical plankton tows collected with a 75 μm net), sedimentary organic carbon and the dissolved inorganic carbon. Figure V.3 shows the distribution of these parameters for different sectors of the Gulf system. Note that data from the deep eastern basin of the Upper Estuary are listed separately from that from the shallow portion of the Upper Estuary.

The most significant feature of the δ^{13}C data is the remarkable uniformity of the POC isotope ratio. In the Upper Estuary, the standard deviation of the distribution is only 0.8‰ ($n = 53$). The δ^{13}C value for POC in this region (-24.4‰) is indistinguishable from POC in the two adjacent areas: the eastern basin and the St. Lawrence River. Samples collected in this study from the River were very homogeneous ($\delta = 0.34$‰, $n = 12$). These isotope ratios are consistent with those observed in the river monitoring study cited above (Fig. V.2) at the time of year when river production is an important component of the organic carbon.

The close similarity between Upper Estuary POC and the large particle material from the vertical plankton tows (VPT) (δ^{13}C $= -24.5$‰, $\sigma = 0.9$, $n = 8$) suggests that both the large particles from the VPT samples and the POC are "fresh" organic matter. Based on these data, Tan and Strain (1983) concluded that the POC in the Upper St. Lawrence Estuary most likely had a riverborne source. They also noted that Cardinal and Berard-Therriault (1976) observed high concentrations ($\approx 10^6$ cells L^{-1}) of predominantly freshwater diatoms in the Upper Estuary in September. The one significant brackish water species identified, *Skeletonema subsalum*, was the most abundant species in samples collected in July and September, suggesting that production within the Estuary may also constitute a significant component of the POC at this time of year.

Saguenay Fjord

Isotope ratios for sedimentary organic carbon in the Saguenay Fjord were reported by Tan and Strain (1979b). The δ^{13}C values observed for nine samples located along the entire length of the Fjord show a range of -26.4 to -24.8‰ with a mean of -25.9‰. The sample with the highest δ^{13}C value (-24.8‰) was in the eastern basin near the mouth of the Saguenay Fjord closest to marine organic matter derived from the highly productive Lower Estuary. Eight samples located in the inner western basin farther upstream show a remarkably narrow range of -26.4 to -25.8‰ ($\sigma = 0.2$‰), identical to the expected value for terrestrial organic matter. Combining the isotope data with C/N ratio data once again gives a clearer picture of the sources of carbon in the sediments. Unlike the isotope data, the C/N ratio of organic matter within the sediments shows a wide range (7.6-43, Pocklington and Leonard 1979). This wide range is partially due to the presence of marine organic carbon in the lower reaches of the Saguenay and partially to the ability of the C/N ratio to distinguish between

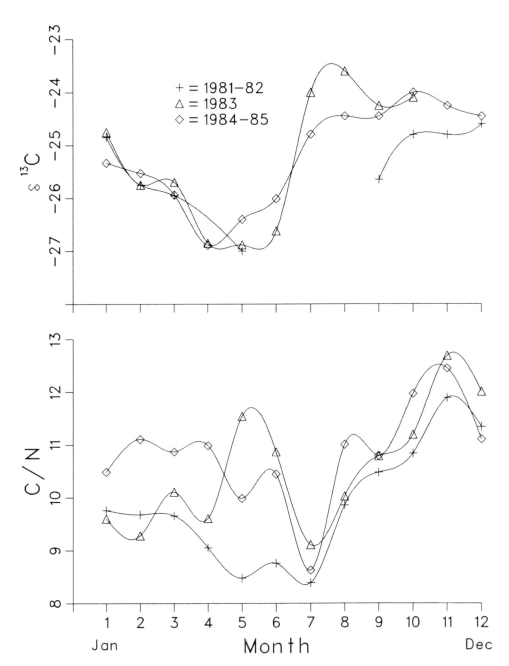

FIG.V.2. Seasonal variability in the $\delta^{13}C$ values and C/N ratios of particulate organic carbon in the St. Lawrence River at Quebec City. (Adapted with permission from Pocklington and Tan 1987.)

different types of terrestrial carbon. The very high C/N ratios observed at the head of the Saguenay result from the high contribution from pulp mill wastes (see Chapter VIII). Similarly high C/N ratios are observed at the head of the Bay of Islands, Newfoundland, near a large paper mill at Corner Brook. The dominance of woody debris in these sediments is confirmed by the detection of high lignin concentrations in these sediments (Pocklington 1976).

Both the isotope and C/N results show that the contribution of marine organic carbon to Saguenay Fjord sediments is small and restricted to the eastern basin. Primary productivity measurements confirm that the amount of carbon produced in Saguenay surface water is small. Therriault and Lacroix (1975) reported that plankton biomass in the top 10-15 m of the water column is lower in the Fjord than in the Estuary, and stated that the high turbidity of Saguenay Fjord waters restricts primary productivity to this unusually shallow surface layer.

Lower St. Lawrence Estuary

Organic carbon in surficial sediments in the Lower Estuary exhibits larger $\delta^{13}C$ gradients than in either the Upper Estuary or the Saguenay Fjord (Fig. V.1). $\delta^{13}C$ values vary between -25.6 and $-21.8°/oo$, with the heavier carbon becoming predominant towards the centre of the Estuary and towards the Gulf of St. Lawrence. Tan and Strain (1979b) ascribed these gradients to the mixing of marine carbon either brought into the Lower Estuary by the saline waters flowing along the Laurentian Channel or produced *in situ*, and terrestrial carbon advected into the Lower Estuary from further upstream. The absence at Rimouski of the very pronounced $\delta^{13}C$ gradient that is evident at the Pointe des Monts section indicates that the influence of marine carbon is confined to approximately the lower half of the Lower Estuary. Since there is no reason why advection of marine carbon by the deep inflowing water of the Laurentian Channel should be limited in this way, the most likely source for the marine carbon is primary production in the overlying surface layer.

Tan and Strain (1979a) first reported that the POC isotope ratio of surface water collected at three stations in the mouth of the St. Lawrence Estuary in mid-August 1974 was heavier than that found in the open Gulf. More extensive sampling in late August, 1979 (Tan and Strain 1983) confirmed the earlier measurements — the heaviest POC in the region was found in surface water of the Lower Estuary and Gaspe current. POC isotope ratios for the St. Lawrence River, Upper Estuary and the surface water ($z < 50$ m) of the open Gulf of St. Lawrence are similar to each other but significantly different from the Lower Estuary and Gaspe means.

The isotope ratios observed in plankton (Fig. V.3) in the Lower Estuary exhibit very similar trends to those observed for POC. The similarity of the isotope ratios for POC and plankton is evidence that the bulk of the POC is composed of living or recently living material. These heavy $\delta^{13}C$ values may be a result of unusually high carbon demand due to the high productivity of the area. Such demand can cause elevated ^{13}C levels in the plankton by reducing the fractionation between inorganic and organic carbon during photosynthesis (e.g. Deuser 1970). (Measurement of the inorganic carbon isotope ratio on samples also collected on cruise 79-024 showed that the isotopic differences between the plankton and inorganic carbon were indeed smaller in the Lower Estuary and Gaspe regions than elsewhere in the Gulf — Tan and Strain 1983.) Alternatively, a change in the dominant planktonic species with an associated shift in isotopic fractionation behavior (Wong and Sackett 1978) could also result in the heavy organic carbon. Further work is required to distinguish between these possible causes. Tan and Strain (1983) concluded that exchanges of POC between the Upper Estuary, the Lower Estuary and Gaspe, and the Gulf were not important in summer. Pocklington (1985a) has recently shown that the Estuary is a source of organic carbon to the Gulf in winter.

Gulf of St. Lawrence

Very little variation in the isotope ratio of organic carbon is observed in sediments in offshore areas of the Gulf. The mean value at 13 stations, excluding the area to the west of Newfoundland's Great Northern Peninsula, is $-22.4 \pm 0.2°/oo$. These carbon isotope ratios are very similar to those at offshore stations outside Cabot Strait (Tan and Strain 1979b). Such uniformity suggests a large and isotopically constant source material derived from plankton production. In addition, this distribution shows that little of the terrestrial organic carbon input to the St. Lawrence Estuary reaches the open

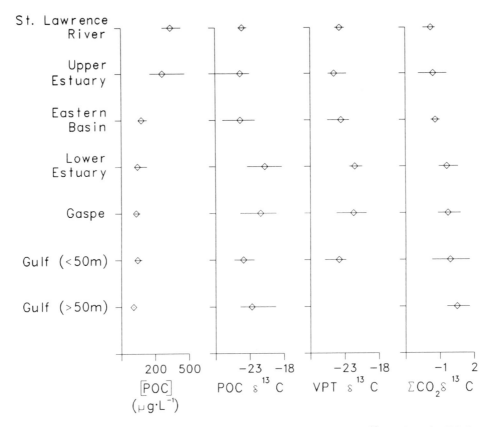

FIG.V.3. Particulate organic carbon concentrations ([POC]), and δ^{13}C values for POC, plankton collected by vertical plankton tows (VPT), and total dissolved CO_2 (ΣCO_2) in different sectors of the Gulf of St. Lawrence. Error bars shown are $\pm$ 1 σ from the mean. Error bars for [POC] are assymetric due to the lognormal distribution found for this parameter (Adapted with permission from Tan and Strain, 1983.)

Gulf. Terrestrial carbon is observed locally nearshore (e.g. the Bay of Islands and Bonne Bay). Rashid and Reinson (1979) reported a gradient from terrestrial (−27‰) to marine (−22‰) carbon in the Miramichi Estuary, very similar to that reported by Tan and Strain (1979b) for the St. Lawrence Estuary. (Rashid and Reinson, working prior to the availability of direct measurements on the isotope composition of marine sediments in the area, chose an inappropriate value for the marine end member in their calculation of the terrestrial carbon concentration in the Miramichi.) Slightly lighter carbon was also found off the west coast of Newfoundland. Since this region is one of comparatively low productivity (see Chapter III), it is possible that inputs of terrestrial matter can be detected farther offshore than elsewhere in the region.

Some limited data is available on the variation of POC isotope ratios with depth in the Gulf of St. Lawrence (Tan and Strain 1979a). The average δ^{13}C value in deep water in the Laurentian Channel (−26.2‰) was lower than that observed both for surface POC (−24.9‰) and in sediments (−22.4‰). In addition, at five of six stations the concentrations and isotope ratios of POC in a bottom water sample were higher than those observed in a deep sample farther from the bottom, suggesting that the bottom water sample had been influenced by resuspension. Tan and Strain suggested that organic debris in the sediments and benthic boundary layer is less degraded than POC in the deep water.

Inorganic Carbon Isotopes

Introduction

The isotope composition of total dissolved inorganic carbon in marine and fresh-water environments has been studied because of its role in determining the isotopic composition of organic matter and its application to the study of various natural processes. In coastal environments, the isotopic composition of total dissolved inorganic carbon has been used: (1) to estimate the relative contribution of organic and carbonate-derived carbon to the total CO_2 concentration in deep water, (2) to examine the pathways of carbon and (3) to study the mixing of water masses.

The procedures used for sampling, extractions, and $^{13}C/^{12}C$ analysis of total dissolved CO_2 in all the studies discussed below are detailed in Tan et al. (1973). The term "total inorganic carbon", or total CO_2 (ΣCO_2), refers to CO_2 produced by the acid extraction of all the dissolved carbonate species: H_2CO_3, HCO_3^-, and $CO_3^=$. The overall analytical precision determined by repeated analysis of aliquots of sea water samples indicates a standard deviation of $0.30°/oo$ for $\delta^{13}C$ and 4 % for total CO_2 concentrations.

$\delta^{13}C$ Variations in Surface Waters

The isotopic composition of the total CO_2 in surface water samples (1 m) from the Gulf of St. Lawrence has been determined as part of several studies (Tan and Walton 1978; Strain and Tan 1979; Tan and Strain 1983). Table V.1 lists the ranges for $\delta^{13}C$ and total CO_2 concentrations reported in these studies. In comparing the entries in the Table, the salinity ranges must be considered — the different cruises sampled different sectors of the mixing zone (see Appendix). Considerable variation occurs in the freshwaters feeding the Gulf of St. Lawrence. Inorganic carbon in the St. Lawrence River (cruise 76-006) has a much higher isotope ratio and concentration ($-4.2°/oo$, 0.88 mM) than that in the Saguenay River measured on the same cruise ($-10.9°/oo$, 0.16 mM). The higher values found in the St. Lawrence River reflect the greater abundance of marine carbonate rocks in the Great Lakes drainage basin; the Saguenay River drains an area dominated by the Precambrian shield (see Chapter I). ΣCO_2 produced partially from the dissolution of marine carbonate rocks will be closer to their isotopic composition (-2 to $+2°/oo$) than CO_2 produced predominantly from the oxidation of isotopically very light terrestrial organic matter ($\approx -26°/oo$).

TABLE V.1. Isotope Ratios and Concentrations for ΣCO_2 in the Gulf. Values listed are for the freshest and most saline samples collected on each cruise.

Region (Cruise)	Season	Salinity	$\delta^{13}C$ ($°/oo$)	ΣCO_2 (mM)
Lower Estuary + open Gulf (73-012)	Apr.-May	23.0	− 1.09	1.70
		31.8	+ 2.75	2.02
Open Gulf (74-028)	July-Aug.	28.8	+ 1.79	1.85
		31.0	+ 2.30	1.96
Upper Estuary (76-006)	Apr.-May	0.0	− 4.2	0.88
		24	− 1.6	1.68
Saguenay Fjord (76-006)	Apr.-May	0.4	− 10.9	0.16
		18	− 0.4	1.35
Estuary + Gulf (79-024)	August	0.0	− 1.8	1.28
		30.9	+ 0.5	2.00

In surface waters of the open Gulf, isotope ratios and concentrations exhibit much smaller ranges (+0.5 to +2.75‰ and 1.96 to 2.02 mM) than in freshwater. These $\delta^{13}C$ values are similar to those reported for the surface waters in the central oceans (+2.2‰, Kroopnick 1974) and to that observed at GEOSECS station II in the North Atlantic (+1.9‰, Kroopnick et al. 1972). This close similarity exists because the ΣCO_2 in waters of the open Gulf are not influenced by freshwater inputs and the ΣCO_2 in all these surface waters approaches isotopic equilibrium with atmospheric CO_2.

The mixing behaviour of the $^{13}C/^{12}C$ ratio of total dissolved inorganic CO_2 in river and sea waters in an estuarine environment is a subject of some interest. A study of this behaviour provides information concerning the processes that govern the distributions of $^{13}C/^{12}C$ in such an environment. A knowledge of the $^{13}C/^{12}C$ ratio and its relationship to the salinity and $^{18}O/^{16}O$ ratio in water in modern nearshore environments is essential in understanding and interpreting the isotope data obtained from fossil carbonate shells deposited in ancient nearshore environments (Mook 1971).

Strain and Tan (1979) described a mixing model for the inorganic carbon isotope ratio in surface waters from the Saguenay Fjord and the Upper St. Lawrence Estuary. Since both the ΣCO_2 concentration and the isotope ratio differ between the fresh and marine waters which mix in these areas, conservative behaviour results in a non-linear $\delta^{13}C$ versus salinity relationship. Figure V.4 compares the predicted behaviour for $\delta^{13}C$ with observations for both the Upper St. Lawrence Estuary and the Saguenay Fjord. In the Saguenay, model and observations agree very well, indicating that both the isotope ratio and concentration of the dissolved CO_2 behave conservatively. It implies that any processes involving uptake and release of CO_2 in the surface water such as photosynthesis, oxidation of organic matter and isotopic exchange with atmospheric CO_2 must occur at rates negligible compared to the residence time of surface water in the Saguenay.

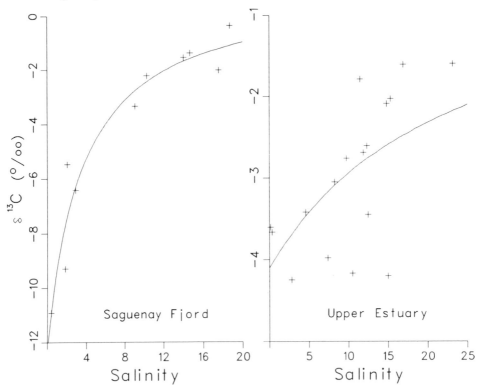

FIG.V.4. Comparison between the $\delta^{13}C$ values predicted for ΣCO_2 by a conservative mixing model (solid lines) and observations (+'s) for the Saguenay Fjord and the Upper St. Lawrence Estuary (adapted from Strain and Tan 1979, with permission).

The situation in the Upper St. Lawrence Estuary is very different to that in the Saguenay Fjord — the model cannot predict the observed $\delta^{13}C$ to better than $\approx \pm 1.0\%$. The deviation from conservative behaviour probably results from a combination of several factors, the most important of which may be the complex nature of the circulation in the Upper St. Lawrence Estuary. It is possible to envision many different processes involving the dissolved CO_2 in the setting of this circulation. For example, suspended organic matter ($\delta^{13}C = -24\%$) which is recirculated within the turbidity zone (see Chapter II) would have a greater chance of being oxidized than if it only passed through the mixing zone once, as is the case in the Saguenay Fjord. (These samples were collected in April, a time of year when in situ processes contribute the least to the organic carbon pool.) The consequence of this oxidation would be to lower the carbon isotope ratio and increase the concentration of the dissolved CO_2.

The isotope results observed for the Saguenay Fjord and the Upper St. Lawrence Estuary have important implications regarding the use of carbon and oxygen isotope ratios of fossil shell carbonates for the estimation of paleotemperatures and paleosalinities in nearshore areas. In such studies (e.g. Mook 1971), the oxygen isotope composition of the shell carbonate is used as an analog of salinity (see below). The $\delta^{13}C$ versus $\delta^{18}O$ relationship is extrapolated to open ocean conditions to yield the average growth temperature of the organism. If the ancient estuary had inorganic carbon behaviour similar to that of the Saguenay, such a temperature determination would only be accurate if data were available to properly describe the curvature of the relationship (i.e. $\delta^{13}C$ vs S or $\delta^{13}C$ vs $\delta^{18}O$). If, on the other hand, the ancient estuary had shown as much scatter as the Upper Estuary, then the resulting temperatures would have large uncertainties. In either case, the errors in temperature can be large: an error of $\approx 1\%$ in determining the isotope ratio translates into an error of $\approx 4°C$ in temperature.

Variation of $\delta^{13}C$ Values with Depth

The variations of the isotope ratio of the ΣCO_2 with depth were examined at a number of selected stations by Tan and Walton (1978). Figure V.5 presents typical $\delta^{13}C$ profiles at two deep stations within the Gulf together with corresponding temperature and dissolved oxygen data. $\delta^{13}C$ values generally decrease progressively with depth from the surface to deep water with an average change of 2.0%. Except for the presence of oxygen maxima near but not at the surface at some stations, the dissolved oxygen concentrations show a similar decrease with depth.

$\delta^{13}C$ values in the surface waters are higher than those at depth because of isotopic exchange with atmospheric CO_2 and also because photosynthetic processes preferentially remove ^{12}C. With increasing depth the $\delta^{13}C$ values decrease because of the decomposition of organic matter which produces ^{13}C depleted CO_2. The CO_2 produced in the decomposition process will have a $\delta^{13}C$ value similar to its organic precursor, which, as noted in a previous section, is about $24-26\%$ lower than the inorganic carbon. These changes in the $\delta^{13}C$ of the inorganic carbon reservoir occur in parallel to those observed in nutrient concentrations with depth (see Chapter III).

During 1974 it was observed that the presence of an oxygen maximum was not accompanied by a corresponding maximum in the $\delta^{13}C$ of the total CO_2 at all stations. If photosynthetic activity is the cause of the oxygen maximum during the summer, one should have observed a corresponding $\delta^{13}C$ maximum. This observation provides independent support to the suggestion of d'Anglejan and Dunbar (1968) that the presence of an oxygen maximum in the Gulf of St. Lawrence during the summer is caused by the seasonal heating of the winter mixed layer rather than by photosynthetic activity.

Some idea of the variability of $\delta^{13}C$ values with depth in 1973 in the St. Lawrence Estuary can be seen in the data from stations 58 and 71 (Fig. V.6). In contrast to those stations located in the Gulf of St. Lawrence, low $\delta^{13}C$ values are found for both the surface and deep waters in the Estuary reflecting the influence of outflowing freshwater from the St. Lawrence River at the surface and the inflow of deep water along the Laurentian channel.

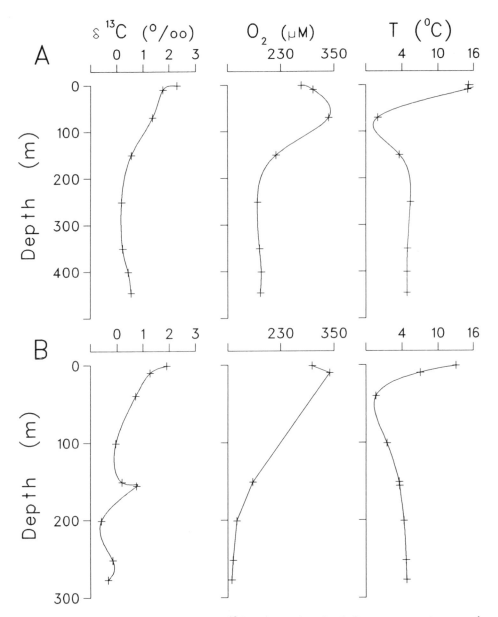

FIG.V.5. Vertical profiles for ΣCO_2 $\delta^{13}C$ values, dissolved O_2 concentrations, and temperatures for two stations from the open Gulf of St. Lawrence. A. Station 23, Cabot Strait. B. Station 63, Pointe des Monts. (redrawn with permission from Tan and Walton 1978).

The carbon isotope ratio of the inorganic carbon can also be used to help quantify the regeneration of organic matter in the deep water of the Laurentian Channel as it flows from Cabot Strait to the mouth of the Saguenay Fjord. Regeneration processes are considered in detail in Chapter III.

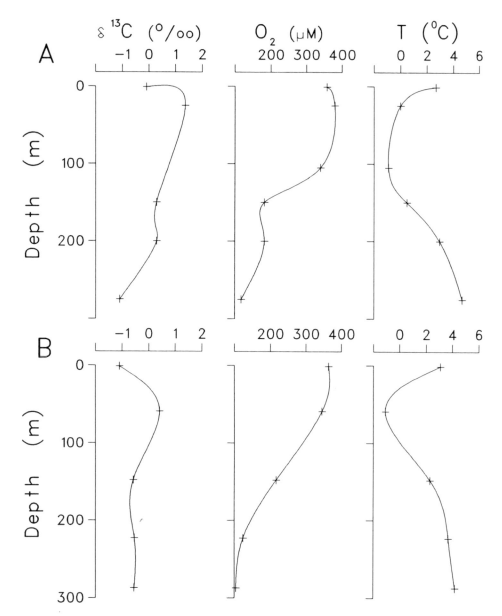

FIG.V.6. Vertical profiles for δCO_2 $\delta^{13}C$ values, dissolved O_2 concentrations, and temperatures for two stations from the Lower St. Lawrence Estuary. A. Station 58, Pointe des Monts. B. Station 71, off Rimouski.

Oxygen Isotopes

Introduction

Estuarine mixing of river and ocean waters has been a subject of considerable interest and study. Salinity trends and variations in concentrations of dissolved constituents of assumed conservative behaviour have been widely used to estimate the degree of mixing. It has been suggested that the application of oxygen isotope variations might be a valuable tracer technique for these mixing processes (Boyle et al. 1974). Such a suggestion stems from the fact that ocean waters are substantially enriched in ^{18}O

in comparison with freshwaters from the continents (Epstein and Mayeda 1953). Thus when two parcels of water differing in salinity and ^{18}O concentration mix, the resulting mixture will have intermediate ^{18}O values determined by the relative proportions of the contributions. The ^{18}O approach has a theoretical advantage in that it directly measures the mixing of the H_2O component of the estuarine fluid. In practice, its application is limited by analytical considerations — the analytical precision of $^{18}O/^{16}O$ measurements is much poorer than that of salinity (by $\approx 100:1$) and the measurement of isotope ratios is very time consuming and requires expensive equipment. Therefore the value of stable isotope measurements is found in applications in which they can provide information not available by simpler means.

^{18}O Values of Surface Waters

A study of the variation of $^{18}O/^{16}O$ ratios in waters collected at selected stations and depths in the St. Lawrence Estuary and the Saguenay Fjord was conducted for the purpose of assessing possible applications of oxygen isotope measurements in the Gulf of St. Lawrence. The water samples were analyzed for $^{18}O/^{16}O$ ratios according to the procedure described by Epstein and Mayeda (1953), and the results are expressed in the usual $\delta^{18}O$ notation with respect to Standard Mean Ocean Water (SMOW). This technique determines the isotopic composition of the oxygen in the water molecules of the fluid. The overall analytical precision, as determined by repeated analysis of aliquots of sea water samples, is $0.13^{0}/_{00}$.

A plot of the $\delta^{18}O$ data against salinity for surface waters in the St. Lawrence Estuary and the Saguenay Fjord is presented in Fig. V.7. The mixing line on the figure joins the two major components for mixing in the St. Lawrence Estuary: St. Lawrence River water with a mean $\delta^{18}O$ value of $-10.3^{0}/_{00}$ and a salinity of 0 and the deep warm water located in the Laurentian Channel of the lower Estuary which has a mean $\delta^{18}O$ value of $+0.15^{0}/_{00}$ and a salinity of 34.6.

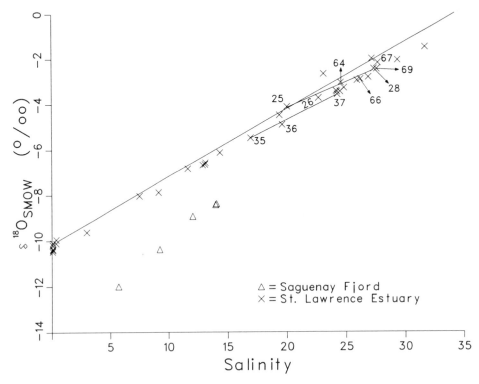

FIG.V.7. $\delta^{18}O$ – salinity relationship for surface waters in the St. Lawrence Estuary and the Saguenay Fjord. Stations identified on the plot are discussed in the text.

A number of interesting aspects of the oxygen isotope distribution are apparent on Fig. V.7. Surface samples from the Upper Estuary (x's at salinities < 15) lie very close to the mixing line. Surface samples from the Saguenay Fjord (triangles) are very clearly not on this mixing line. Extrapolation of the Saguenay samples to zero salinity gives a $\delta^{18}O$ freshwater value of $-14.0°/oo$. The different isotopic composition of freshwaters in the St. Lawrence and Saguenay rivers is due to the well known decrease of ^{18}O content in precipitation with increasing latitude (e.g. Hage et al. 1975) — the Saguenay drainage basin is further north than most of that of the St. Lawrence.

This isotopic difference between the Saguenay and St. Lawrence freshwaters makes it possible to distinguish between them in some parts of the Estuary, a distinction that cannot be made on the basis of salinity alone. Stations 35-37 of cruise 74-006 were a cross-Estuary section ≈ 10 km upstream of the mouth of the Saguenay. Surface waters in this section fall below the mixing line in Fig. V.7, clearly showing the influence of the Saguenay discharge (the presence of Saguenay water 10 km upstream of the Saguenay's mouth is within the scope of the tidal excursions in this part of the Estuary — see Chapter I). Stations 25-28 formed another cross-section of the Estuary ≈ 15 km downstream of the mouth of the Saguenay (see the Appendix for the exact positions of these stations). The influence of Saguenay freshwaters in offsetting the points from the mixing line, especially at station 28 on the north side of the Estuary, is again apparent in Fig. V.7.

In theory, it is possible to calculate the amount of water from each source (Saguenay freshwater, St. Lawrence freshwater, and Laurentian Channel seawater) based on the position of the point on the $\delta^{18}O$-S diagram. In practice, this calculation may apply only to the area of the Estuary close to the mouth of the Saguenay, where the signal is least likely to be overwhelmed by the larger St. Lawrence discharge (≈ 9 × the discharge of the Saguenay — see Chapter I), and would require careful determination of the isotope ratios so that the analytical precision would not produce unacceptable errors in the final result.

Deviations from the mixing line are also found for stations 64, 66, 67, and 69 farther downstream in the Lower Estuary. These deviations probably reflect the influence both of the Saguenay and of significant local contributions from other rivers draining northern areas such as the Aux Outardes and the Manicouagan which flow into the lower part of the Lower Estuary. Characterization of the isotope ratios of freshwaters from these other rivers would be necessary prior to calculating freshwater components at these stations. Should these rivers be isotopically similar to the Saguenay, the calculation would yield the relative contributions of the St. Lawrence and the "northern rivers". If the isotopic compositions of these other rivers differ significantly from that of the Saguenay, then the isotope — salinity data could not produce unambiguous source water information, although it might be possible to place some constraints on relative freshwater contributions.

$\delta^{18}O$ Distributions in the Gulf of St. Lawrence

As discussed in the previous section, it is possible to characterize freshwaters by the determination of their oxygen isotope composition. In a similar way, it is also possible to characterize more saline waters by a determination of the isotopic composition of their apparent freshwater component. Figure V.8 gives several examples of $\delta^{18}O$-S plots. Figure V.8 A shows the simplest case: (i) mixing in the Saguenay Fjord occurs between two well-defined end-members: freshwater from the Saguenay River and the deep, saline water in the Fjord and (ii) samples are available over virtually the entire salinity range, which produces a highly precise estimate for the isotopic composition of the freshwater "component" (i.e. Saguenay River water).

Figure V.8 B shows a $\delta^{18}O$-S plot for samples taken throughout the water column in the Laurentian Channel east of Cabot Strait (i.e. the source water for the deep water in the Gulf). The salinities of all these samples fall in a narrow range between 31 and 35, so that the extrapolation to zero salinity is much less precise than in the Saguenay.

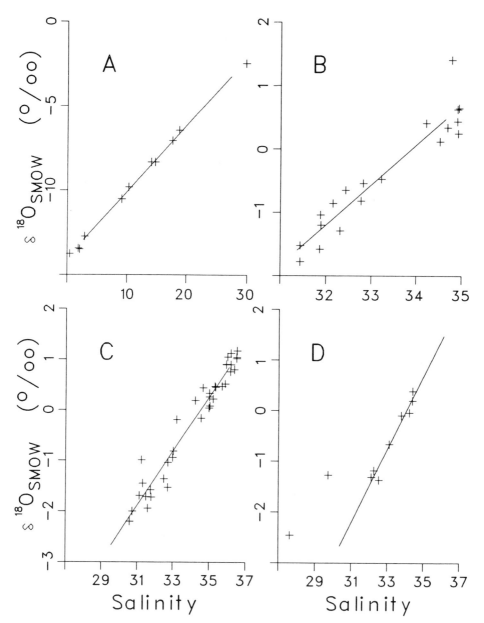

FIG.V.8. Some typical $\delta^{18}O$ – salinity plots for several environments in and near the Gulf of St. Lawrence. A. The Saguenay Fjord. B. Laurentian Channel, east of Cabot Strait. C. Scotian Shelf. D. Near the mouth of the St. Lawrence Estuary.

The $\delta^{18}O$ value of the freshwater component determined from this extrapolation is $-20.8\%o$. This value is typical of deep waters found at much higher latitudes in the Northern Labrador Sea and Davis Strait (Tan and Strain 1980), and is consistent with the known high latitude source of sub-surface open ocean waters.

Figure V.8 C shows the $\delta^{18}O$-S relationship for samples collected on a section across the Scotian Shelf southeast from Halifax in October, 1975. The apparent freshwater component has a $\delta^{18}O$ value of $-18.5\%o$. It is tempting to suggest that this indicates that the water on the Scotian Shelf contains freshwaters from both the Laurentian Channel water entering the Gulf and the local Gulf inputs, but the isotope ratios of the freshwaters implied by Fig. V.8 B and C are not significantly different.

Finally, Fig. V.8 D shows data from one station in the mouth of the St. Lawrence Estuary. The points at salinities > 31 (below the halocline) fit closely to a $\delta^{18}O$-S line that the points with salinities < 31 do not fit. This is a clear demonstration that the freshwater components of surface and deep waters at this station differ. This difference raises the possibility of using oxygen isotope measurements to help describe mixing processes both inside the Gulf and in the outflow from the Gulf on the Scotian Shelf. Within the Gulf, it may be possible to follow the dilution of local Gulf freshwater inputs by the northern freshwater component in deep Gulf water by mapping the isotopic composition of the freshwater component in surface samples. Since the freshwater isotope ratio must be based on samples covering a very small salinity range, its determination would not be trivial. Great care would be necessary in sample selection and obtaining the best possible isotope data.

The isotopic makeup of the freshwater component in the outflow from the Gulf through Cabot Strait is of interest in view of several studies of nearshore dynamics off eastern North America. For example, Chapman et al. (1986) used oxygen isotope data to follow mixing between shelf water (which they believed to be derived from the Scotian Shelf) and slope water in the mid-Atlantic Bight; Fairbanks (1982) reported that he had detected sea-ice meltwater from the Gulf of St. Lawrence in the New York Bight. A study into the isotope ratios of freshwater components in the Gulf of St. Lawrence and on the Scotian Shelf is currently underway.

Oxygen Isotope Fractionation between Sea Ice and Seawater

The behaviour of deuterium during the freezing of water has been examined theoretically by Weston (1955). His results suggested that when water freezes under equilibrium conditions, the deuterium content of the ice should be about $20°/oo$ higher than that of the liquid water. Friedman et al. (1964) have analyzed the deuterium content of sea ice and underlying coexisting seawater from various locations and found that the ice contained an average of $17°/oo$ more deuterium than the seawater, in good agreement with the theoretical deductions. Because the ^{18}O content in water is about a factor of 10 less than that of deuterium, it is expected that ice will contain about $2°/oo$ more ^{18}O than the water from which it freezes. This degree of enrichment in ice relative to water has been confirmed experimentally by O'Neil (1968) for distilled water.

Oxygen isotopic fractionation between sea ice and seawater in samples collected from various locations in the Gulf of St. Lawrence was examined by Tan and Fraser (1976). Samples were collected in late winter, 1974, according to the procedures described in Dunbar (1973). Station locations are shown in Fig. V.9. The $\delta^{18}O$ results for surface seawater, sea ice, and snow are presented in Table V.2.

The longest ice core was collected at station 33 (81 cm). Three sections of sea ice, from 0 to 61 cm, had essentially identical $\delta^{18}O$ values (0 to $+0.2°/oo$). The $\delta^{18}O$ value of the underlying seawater was $-1.9°/oo$. Thus the sea ice was enriched in ^{18}O by $\approx 2.0°/oo$. The isotope ratio of the uppermost section of the core was $-2.1°/oo$, lower than the three lower sections. This reduction is due to mixing of "normal" sea ice at $\approx 0°/oo$ with the overlying snow cover which had a $\delta^{18}O$ value of $-8.7°/oo$.

The next longest ice core was 61 cm (station 28). The same pattern in the $\delta^{18}O$ distribution is repeated. The snow cover had a $\delta^{18}O$ value of $-10.9°/oo$ and the sea ice immediately beneath a value of $-6.8°/oo$. Once again the low isotope ratio in the upper ice section is due to mixing of snow into the ice. The slightly depressed $\delta^{18}O$ value in the 20-41 cm section ($-0.5°/oo$) at this station may indicate that some snow, snow melt or rain has penetrated further into the ice than at station 33. Again the enrichment of ^{18}O in sea ice with respect to seawater, using the bottom section of the core as representative of "uncontaminated" sea ice, was $\approx 2.0°/oo$.

Somewhat smaller fractionation values are observed in the remaining two cores ($1.4°/oo$ at station 27 and $1.0°/oo$ at station 32). These ice floes were thinner than those discussed above, with the result that mixing of precipitation with the ice would be more likely to penetrate to the 0-20 cm sections. At station 32, greater perturbation of the

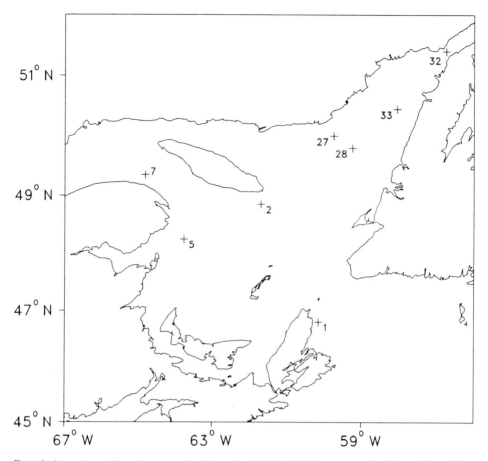

FIG. 9 Location of sea-ice samples from the Gulf of St. Lawrence (redrawn from Tan and Fraser 1976).

sea ice isotope ratio would be expected because of the very low $\delta^{18}O$ value of the snow cover (-18.0%).

In addition to the ice core data, Table V.2 also lists isotope ratios for slush ice and surface sea water, and their differences, at four other stations. The isotopic enrichment in sea ice relative to seawater ranged from 1.3 to 2.3% with a mean of 1.8%. This value is the same as the mean obtained from the ice cores, excluding the sample from Station 32.

Although the thickness of the sections in the ice cores did not allow an exact determination of the isotopic fractionation between sea ice and seawater, the observed fractionation is in excellent agreement with both theoretical expectations and the distilled water experimental data (O'Neil 1968). Most of the salt in seawater is rejected during sea ice formation — Tan and Strain (1980) reported an average salinity of 6.1 for 104 samples of mature first year ice. Since freshwater from the melting of sea ice is isotopically much heavier than the normal freshwater component of seawater with salinities of ≈ 30, sea-ice meltwater can be distinguished from other freshwaters on a $\delta^{18}O$-S diagram. Meltwater measurements have not been made in the Gulf of St. Lawrence, but the expected isotopic signature of meltwater from the Gulf is the basis for its reported detection in the New York Bight (Fairbanks 1982). Using isotope ratios to evaluate the importance of sea ice meltwater in the Gulf outflow is one aspect of the study underway at BIO.

TABLE V.2. $\delta^{18}O$ values for sea ice and surface seawater in the Gulf of St. Lawrence. Station locations are shown in Fig. V.9.

Station	Sample Type	$\delta^{18}O_{smow}$ (°/oo)	$\delta_{ice} - \delta_{water}$ (°/oo)
27	Ice core (20 - 41 cm)	− 4.8	
	Ice core (0 - 20 cm)	− 0.1	
	Seawater	− 1.5	1.4
28	Snow	− 10.9	
	Ice core (41 - 61 cm)	− 6.8	
	Ice core (20 - 41 cm)	− 0.1	
	Ice core (0 - 20 cm)	+ 0.5	
	Seawater	− 1.5	2.0
32	Snow	− 18.0	
	Ice core (20 - 41 cm)	− 8.7	
	Ice core (0 - 20 cm)	− 0.9	
	Seawater	− 1.9	1.0
33	Snow	− 8.7	
	Ice core (61 - 81 cm)	− 2.1	
	Ice core (41 - 61 cm)	0.0	
	Ice core (20 - 41 cm)	+ 0.2	
	Ice core (0 - 20 cm)	+ 0.1	
	Seawater	− 1.9	2.0
1	Slush ice	+ 0.5	
	Seawater	− 1.8	2.3
2	Slush ice	− 0.2	
	Seawater	− 1.5	1.3
5	Slush ice	+ 0.2	
	Seawater	− 1.7	1.9
7	Slush ice	+ 0.1	
	Seawater	− 1.4	1.5

References

BOYLE, E. A., R. COLLIER, A. T. DENGLER, J. M. EDMOND, A. C. NG, AND R. F. STALLARD. 1974. On the chemical mass-balance in estuaries. Geochim. Cosmochim. Acta 38: 1719-1728.

CARDINAL, A., AND L. BERARD-THERRIAULT. 1976. Le phytoplancton de l'estuaire moyen du Saint Laurent en amont de l'Ile-aux-Coudres (Quebec). Int. Rev. Gesatem Hydrobiol. 61: 639-648.

CHAPMAN, D. C., J. S. BARTH, R. C. BEARDSLEY, AND R. G. FAIRBANKS. 1986. On the continuity of mean flow between the Scotian Shelf and the Middle Atlantic Bight. J. Phys. Oceanogr. 16: 758-772.

CRAIG, H. 1953. The geochemistry of the stable carbon isotopes. Geochim. Cosmochim. Acta 3: 53-92.

D'ANGLEJAN, B. F., AND M. J. DUNBAR. 1968. Some observations of oxygen, pH and total alkalinity in the Gulf of St. Lawrence, 1966, 1967, 1968. McGill Univ. Mar. Sci. Cent. Ms. Rep. 7: 50 p.

DEUSER, W. G. 1970. Isotopic evidence for diminishing supply of available carbon during diatom bloom in the Black Sea. Nature (Lond.) 225: 1069-1071.

DUNBAR, M. J. 1973. Chlorophyll and nutrient measurements in sea ice, Gulf of St. Lawrence, p. 106-127. *In* Proceedings of Workshop of Physical Sciences in the Gulf and Estuary of St. Lawrence, held at Université du Québec à Rimouski, October 11-12, 1973.

ECKLEMANN, W. R., W. S. BROECKER, D. W. WHITLOCK, AND J. R. ALLSUP. 1962. Implications of carbon isotopic composition of total organic carbon of some recent sediments and ancient oils. Am. Assoc. Petr. Geol. Bull. 46: 699-704.

EPSTEIN, S., AND T. MAYEDA. 1953. Variation of ^{18}O content of waters from natural sources. Geochim. Cosmochim. Acta 4: 213-224.

FAIRBANKS, R. G. 1982. The origin of continental shelf and slope water in the New York Bight and the Gulf of Maine: Evidence from $H_2{}^{18}O/H_2{}^{16}O$ ratio measurements. J. Geophys. Res. 87: 5796-5808.

FRIEDMAN, I., A. C. REDFIELD, B. SCHOEN, AND J. HARRIS. 1964. The variation of the deuterium content of natural waters in the hydrologic cycle. Rev. Geophys. 2: 177-224.

GARDNER, W. S., AND D. W. MENZEL. 1974. Phenolic aldehydes as indicators of terrestrially derived organic matter in the sea. Geochim. Cosmochim. Acta 38: 813-822.

HAGE, K. D., J. GRAY, AND J. C. LINTON. 1975. Isotopes in precipitation in northwestern North America. Monthly Weather Rev. 103: 958-966.

HEDGES, J. I., AND P. L. PARKER. 1976. Land-derived organic matter in surface sediments from the Gulf of Mexico. Geochim. Cosmochim. Acta 40: 1019-1029.

KROOPNICK, P. 1974. Correlations between ^{13}C and ΣCO_2 in surface waters and atmospheric CO_2. Earth Planet. Sci. Lett. 22: 397-403.

KROONPNICK, P., R. F. WEISS, AND H. CRAIG. 1972. Total CO_2, ^{13}C and dissolved oxygen - ^{18}O at Geosecs II in the North Atlantic. Earth Planet. Sci. Lett. 16: 103-110.

MOOK, W. G. 1971. Paleotemperatures and chlorinities from stable carbon and oxygen isotopes in shell carbonate. Paleogeography, Paleoclimat. Paleoecol. 9: 245-263.

NEWMAN J. W., P. L. PARKER, AND E. W. BEHRENS. 1973. Organic carbon isotope ratios in Quarternary cores from the Gulf of Mexico. Geochim. Cosmochim. Acta 37: 225-238.

NISSENBAUM, A. 1974. Deuterium content of humic acids from marine and non-marine environments. Mar. Chem. 2: 59-63.

NISSENBAUM, A., AND I. R. KAPLAN. 1972. Chemical and isotopic evidence for the *in situ* origin of marine humic substances. Limnol. Oceanogr. 17: 570-582.

O'NEIL, J. R. 1968. Hydrogen and oxygen isotope fractionation between ice and water. J. Phys. Chem. 72: 3683-3684.

PARKER, P. L., E. W. BEHRENS, J. A. CALDER, AND D. J. SHULTZ. 1972. Stable carbon isotope ratio variations in the organic carbon from Gulf of Mexico sediments. Contribution in Marine Science, University of Texas 16: 139-147.

POCKLINGTON, R. 1973. Organic carbon and nitrogen in sediments and particulate matter from the Gulf of St. Lawrence. Bedford Inst. Oceanogr. Rep. BI-R-73-8: 16 p.

　　　1975. Carbon, hydrogen, nitrogen and lignin determinations on sediments from the Gulf of St. Lawrence and adjacent waters. Bedford Inst. Oceanogr. Rep. BI-R-75-6: 12 p.

　　　1976. Terrigenous organic matter in surface sediments from the Gulf of St. Lawrence. J. Fish. Res. Board Can. 33: 93-97.

　　　1985a. Organic matter in the Gulf of St. Lawrence in winter. Can. J. Fish. Aquat. Sci. 42: 1556-1561.

　　　1985b. The contribution of organic matter by the St. Lawrence River to the Gulf of St. Lawrence, 1981-1983. Mitt. Geol. Palaont. Inst. Univ. Hamburg 58: 323-329.

POCKLINGTON, R., AND J. D. LEONARD. 1979. Terrigenous organic matter in sediments of the St. Lawrence estuary and the Saguenay Fjord. J. Fish. Res. Board. Can. 36: 1250-1255.

POCKLINGTON, R., AND F. C. TAN. 1983. Organic carbon transport in the St. Lawrence River. Mitt. Geol.-Palaont. Inst. Univ. Hamburg 55: 243-251.

___ 1987. Seasonal and annual variations in the organic matter contributed by the St. Lawrence River to the Gulf of St. Lawrence. Geochim. Cosmochim. Acta 51: 2579-2586.

RASHID, M. A., AND G. E. REINSON. 1979. Organic matter in surficial sediments of the Miramichi Estuary, New Brunswick, Canada. Estuarine Coastal Mar. Sci. 8: 23-36.

SACKETT, W. M. 1964. The depositional history and isotopic organic carbon composition of marine sediments. Mar. Geol. 2: 173-185.

SMITH, B. N., AND S. EPSTEIN. 1971. Two categories of $^{13}C/^{12}C$ ratios for higher plants. Plant Physiol. 47: 380-384.

STRAIN, P. M., AND F. C. TAN. 1979. Carbon and oxygen isotope ratios in the Saguenay Fjord and the St. Lawrence Estuary and their implications for paleoenvironmental studies. Estuarine Coastal Mar. Sci. 8: 119-126.

TAN, F. C. 1988. Discharge and carbon isotope composition of particulate organic carbon from the St. Lawrence River, Canada. Mitt. Geol.Palaont. Inst. Univ. Hamburg (In press)

TAN, F. C., AND W. D. FRASER. 1976. Oxygen isotope studies on sea ice in the Gulf of St. Lawrence. J. Fish. Res. Board Can. 33: 1397-1401.

TAN, F. C., AND P. M. STRAIN. 1979a. Carbon isotope ratios of particulate organic matter in the Gulf of St. Lawrence. J. Fish. Res. Board Can. 36: 678-682.

___ 1979b. Organic carbon isotope ratios in recent sediments in the St. Lawrence Estuary and the Gulf of St. Lawrence. Estuarine Coastal Mar. Sci. 8: 213-225.

___ 1980. The distribution of sea ice meltwater in the eastern Canadian Arctic. J. Geophys. Res. 85: 1925-1932.

___ 1983. Sources, sinks and distribution of organic carbon in the St. Lawrence Estuary, Canada. Geochim. Cosmochim. Acta 47: 125-132.

TAN, F. C., AND A. WALTON. 1978. Stable isotope studies in the Gulf of St. Lawrence, Canada. p. 27-37 In B.W. Robinson [ed.] Stable isotopes in the earth sciences. New Zealand Department of Scientific and Industrial Research Bull. 220, Wellington, N.Z.

TAN, F. C., G. J. PEARSON, AND R. W. WALKER. 1973. Sampling, extraction and $^{13}C/^{12}C$ analysis of total dissolved CO_2 in marine environments. Bedford Inst. Oceanogr. Rep. BI-R-73-16: 17 p.

THERRIAULT, J.-C., AND G. LACROIX. 1975. Penetration of the deep layer of the Saguenay Fjord by surface waters of the St. Lawrence Estuary. J. Fish. Res. Board Can. 32: 2373-2377.

WESTON, R. E. 1955. Hydrogen isotope fractionation between ice and water. Geochim. Cosmochim. Acta 8: 281-284.

WONG, W. W., AND W. M. SACKETT. 1978. Fractionation of stable carbon isotopes by marine phytoplankton. Geochim. Cosmochim. Acta 42: 1809-1815.

CHAPTER VI

Trace Metals in the Water Column

P. A. Yeats

*Marine Chemistry Division, Physical and Chemical Sciences Branch,
Department of Fisheries and Oceans, Bedford Institute of Oceanography,
P.O. Box 1006, Dartmouth, N.S. B2Y 4A2*

Introduction

Rivers are generally identified as the main sources of trace metals to the ocean. Other sources, such as atmospheric fallout or hydrothermal inputs, are also important but the magnitudes of inputs from these sources are more difficult to estimate and their effects are less easily observed than are the effects of rivers. A number of geochemical processes affect the transport of river-derived metals through estuarine and coastal environments. These include deposition and resuspension of metal-containing particles, precipitation or dissolution reactions, and diagenetic release of dissolved metals from sediments. The St. Lawrence Estuary and Gulf of St. Lawrence have provided an unusual natural laboratory for the study of many of these processes.

Metal behavior in estuaries is influenced both by the changes in the physical chemistry of the water — salinity, E_h, pH, etc. — that occur during mixing of river water with seawater, and by the changing concentrations and character of suspended particulate matter (SPM), particularly in the vicinity of estuarine turbidity maxima. Diagenetic processes in the sediments will also influence the metal distribution in the overlying water. The effects of all these processes can be seen in the metal distributions in the St. Lawrence Estuary. Perhaps because of the large water discharge of the St. Lawrence River compared to rivers feeding most of the other estuaries where trace metal studies have been conducted, erratic temporal and spatial variability has been less of a hindrance to interpretation of results from the St. Lawrence than it has been in some other estuaries. In addition, observations in the Lower St. Lawrence Estuary and Gaspé current permit continuation of the study of estuarine processes over much larger time and distance scales than in most other estuaries.

Description of geochemical processes in non-estuarine coastal areas is more difficult than for estuarine regions. Gradients in metal concentrations (or salinity) will be much smaller on the continental shelves than in estuaries, and water movement will not be constrained by the estuarine geometry. Much of the Gulf of St. Lawrence can be considered a typical example of a general coastal environment. However, the Gulf's geographic setting with relatively narrow straits connecting it to the North Atlantic and its reasonably well-understood water circulation regime simplify the study of some chemical processes in comparison with many other coastal environments.

In this chapter, results from the analyses of unfiltered samples will be referred to as total metals while those based on filtered samples will be called dissolved. In both cases the results are reported in units of $\mu g/L$ or ng/L of water. Particulate metals will refer to the analyses of SPM samples and the results will be in units of $\mu g/g$ of SPM. In most cases the dissolved and total results are based on graphite furnace atomic absorption analyses of samples concentrated by solvent extraction and the particulate results on atomic absorption analyses of filter digests.

St. Lawrence River

The St. Lawrence River is the single most important source of trace metals to the Gulf of St. Lawrence. The dissolved, non-detrital (acetic acid leachable) particulate and detrital (non-leachable) particulate metal concentrations in St. Lawrence River water have been measured on a monthly basis over a 2-year period by Yeats and Bewers

(1982). Instantaneous concentrations, measured during estuarine studies, have also been reported by a number of others (Subramanian and d'Anglejan 1976; Bewers and Yeats 1978; Cossa and Poulet 1978; Gobeil et al. 1981 and 1983; Stoffyn-Egli 1982; Campbell and Yeats 1984). These measurements are summarized in Table VI.1. Radionuclide concentrations in the river have been measured by Serodes and Roy (1983).

The time series study of Yeats and Bewers (1982) defined the average concentrations of dissolved and total particulate Al, Mn, Fe, Co, Ni, Cu, Zn, and Cd as well as determining the variability on a monthly time scale. These data cannot provide information regarding shorter time-scale variability. Subramanian and d'Anglejan (1976) reported dissolved Mn, Fe, Co, Ni, and Zn concentrations for July/August 1974. Although their Fe concentration agrees well with Yeats and Bewers' results for the same months of 1974, the Mn, Ni, and Zn concentrations are 2-4 times as high and Co about 70 times higher than those reported by Yeats and Bewers (1982). This discrepancy is probably due to limitations of the flame atomic absorption methods used by Subramanian and d'Anglejan or contamination during their collection, filtration, or analytical procedures.

The metal content of the river SPM has been measured by Cossa and Poulet (1978), Gobeil et al. (1981) and Yeats and Bewers (1982) (Table VI.1). The particulate Fe, Mn, Zn, and Cd data from the three sources agree well and add credence to the various measurements. The particulate Al, Si, Ca, and Mg data of Gobeil et al. are rather lower than the average concentrations reported by Yeats and Bewers (1982). This may reflect seasonal variability since Yeats and Bewers did not investigate the seasonal variability of the major constituents of the SPM.

The two main conclusions that we can draw from these results are that, in general, the levels of both dissolved and particulate metals are lower than in many other rivers and that the month-to-month variability is also quite low. Care must be taken, however, to distinguish between older measurement of trace metals in rivers and more recently reported measurements made using modern "clean" techniques. Compared to the latter, the levels in the St. Lawrence are intermediate. A number of rivers have some-

TABLE VI.1. Trace metal concentrations in the St. Lawrence River.

Metal	Dissolved $\mu g \cdot L^{-1}$	Reference[a]	Particulate $\mu g \cdot g^{-1}$	Reference[a]
Li	2.0 (Aug.)[b]	5		
Al	64	7	51000 (May)	3
			73000	7
Cr	0.7 (Sept.)	1		
Mn	6.2	7	1250 (May)	3
	14 (July/Aug.)	6	1808 (Sept.)	2
			1150	7
Fe	56	7	46000 (May)	3
	60 (July/Aug.)	6	53000	7
Co	0.16	7	3.8[c]	7
	6.5-8.5 (July/Aug.)	6		
Ni	1.5	7	26[c]	7
	3.2-6.9 (July/Aug.)	6		
Cu	2.5	7	71[c]	7
Zn	8.6	7	403 (Sept.)	2
	29.2-32.5 (July/Aug.)	6	305[c]	7
Cd	0.11	7	1.8 (Sept.)	2
			2.8[c]	7
Hg	< 0.003 (June)	4		
Pb			96 (Sept.)	2

a References: (1) Campbell and Yeats 1984; (2) Cossa and Poulet 1978; (3) Gobeil et al. 1981; (4) Gobeil et al. 1983; (5) Stoffyn-Egli 1982; (6) Subramanian and d'Anglejan 1976; (7) Yeats and Bewers 1982.
b Indicates month in which samples were taken. Data from Yeats and Bewers (reference 7) are annual averages.
c Non-detrital fraction

what lower concentrations for many of these metals, notably Ni and Cd (e.g. the Amazon, Boyle et al. 1982; the Gota, Danielsson et al. 1983) and Zn (Shiller and Boyle 1985) but others, such as the Hudson (Klinkhammer and Bender 1981) and the Rhine (Duinker and Nolting 1978), are significantly higher. With the exception of particulate Al, Fe, Mn, and Zn, whose variabilities are related to water flow, the observed variability in neither the dissolved nor the particulate (on a $\mu g \cdot g^{-1}$ of SPM basis) metal concentrations is simply related to water flow or suspended load.

Upper St. Lawrence Estuary

The first study of trace metal distributions in the St. Lawrence Estuary was made by Subramanian and d'Anglejan (1976) who reported dissolved Fe and Mn versus salinity relationships for the Estuary based on a survey in July/August 1974. The Mn/salinity relationship was essentially linear indicating conservative mixing between the river water and saline waters from the Lower Estuary. The Fe/salinity relationship was distinctly curved and indicative of removal of dissolved iron within the Estuary. The shape of the non-linear Fe/salinity curve might suggest that Fe removal was most intense at a salinity of ≈ 17 in the region of a major turbidity front (d'Anglejan and Ingram 1976). The estuarine behaviour identified by Subramanian and d'Anglejan, and in the other studies described in this section, are summarized in Table VI.2.

Subsequent investigators have studied trace metal distributions in the water (Bewers and Yeats 1978) and particulates (Cossa and Poulet 1978; Gobeil et al. 1981). The trends in dissolved Mn (Mn_d) concentrations reported by Bewers and Yeats based on sampling in May 1974 agree with the work of Subramanian and d'Anglejan (1976). Again a more or less linear Mn_d versus salinity relationship is seen. The scatter in the

TABLE VI.2. Summary of metal distributions in the Upper St. Lawrence Estuary.

Metal	Total	Dissolved	Particulate
Mn	maximum in turbidity zone[1]	conservative[1,8]	minimum in turbidity zone[4]; maximum in river, decreasing through Estuary[3]
Fe	maximum in turbidity zone[1]	loss[1,8]	minimum in turbidity zone[4]
Co,Ni,Cu	maximum in turbidity zone[1]	conservative[1]	
Zn	loss[1]		maximum in river, decreasing through Estuary[3]
Cd,Pb			maximum in river, decreasing through Estuary[3]
Li		conservative[7]	
Cr		loss[2]	
Hg	maximum in turbidity zone[5]	constant[5]	
Se		Se^{IV} removed at low salinity[9] Inorganic Se conservative[9] Organic Se constant[9]	
$^{137}Cs, ^{144}Ce$ $^{226}Ra, ^{228}Th$	maximum in turbidity zone[6]		
$^{7}Be, ^{106}Ru$	decreasing concentration with increasing salinity[6]		
^{235}U	midestuarine maximum[6]		

a References: (1) Bewers and Yeats 1978; (2) Campbell and Yeats; (3) Cossa and Poulet 1978; (4) Gobeil et al. 1981; (5) Gobeil et al. 1983; (6) Serodes and Roy 1983; (7) Stoffyn-Egli 1982; (8) Subramanian and d'Anglejan 1976; (9) Takayanagi and Cossa 1985.

results of both studies may hide some non-linearity but any deviations from linearity would be much smaller than those observed for Fe. The Fe distribution in Bewers and Yeats (1978) shows a general non-linear removal relationship except in the Pte. aux Orignaux/Cap aux Oies region. Here, in the region of the turbidity front (see Chapter II) anomalously high Fe_d concentrations are observed. This anomaly is observed for too many samples to be easily explained by analytical problems and remains an unexplained phenomenon. Fe concentrations in unfiltered samples are not elevated. Dissolved Co, Ni and Cu relationships with salinity are all approximately linear.

Total Mn, Fe, and Co distributions (Bewers and Yeats 1978) mostly mimic those of SPM. This is not surprising since the particulate fraction contributes well over half of the total concentration of these metals. Total nickel, copper, and zinc distributions, while still showing similar features, are less analogous to the SPM distributions as would be expected for metals with increasingly important dissolved fractions.

Flux estimates reported in this paper (Bewers and Yeats 1978) show that although essentially no net removal of SPM from the water column occurs within the Upper Estuary, 11 % of the particulate iron input, 38 % of the particulate manganese (Mn_p) input and 41 % of the Co_p are also removed but the Mn_d flux is increased by 15 %. The increase in Mn_d flux will account for part of the loss of Mn_p. Total copper, zinc and cadmium and dissolved Co and Fe fluxes are reduced by 5-44 % but the total nickel flux out of the Upper Estuary exceeds the inputs by 13 %. These fluxes would imply accumulation of metals (except Ni) in the sediments of the Upper Estuary although accumulation will not necessarily occur at other times of the year.

Particulate iron and manganese distributions in the Estuary were described by Gobeil et al. (1981) based on surveys (including tidal stations) in May and November, 1976. They found high Mn_p and Fe_p contents (on the basis of Mn/Al ratios) in SPM in the river that decreased markedly within the upstream region of the turbidity maximum. In the downstream portion of the turbidity maximum, Mn_p concentrations tended to increase again but did not generally reach the levels observed in the river. In the eastern basin, intermediate to high Mn_p concentrations were observed, with slightly higher concentrations in the deep water. Cossa and Poulet (1978) also described the distribution of Mn_p (Sept. 1974) as well as those of Zn_p, Pb_p, and Cd_p. Their Mn_p results are generally similar to those of Gobeil et al. although Cossa and Poulet found higher river concentrations and lower eastern basin concentrations. Zn_p, Pb_p, and Cd_p concentrations were also found to decrease markedly in the upstream part of the Estuary. Bewers and Yeats (1978) found a somewhat different distribution of Mn_p. Their calculated Mn_p/Fe_p ratios would suggest that there is a slight increase in Mn_p (as well as Co_p and Cu_p) from the river to the turbidity maximum. These results for particulate metal concentrations were based on the difference between total and dissolved water samples rather than direct analysis of the SPM. Settling of SPM in the samples during retrieval and subsampling could cause a bias in these results if there is differential settling of the metals. In addition, the riverine Fe_p concentrations found in this survey appear to be somewhat high compared to those found in the 2-year monitoring of the river (Yeats and Bewers 1982). Unusually rapid loss of Fe_p from the river to the Estuary would make the Mn_p observations consistent with those of Gobeil et al. and Cossa and Poulet.

A fairly consistent picture of the manganese distribution results from these studies. Particulate manganese content of the SPM in the river is relatively high, decreasing to low levels near the head of the turbidity maximum and possibly increasing somewhat in the downstream and deeper parts of the turbidity maximum. The Mn content of the SPM leaving the Estuary in the surface outflow is generally lower than that in both the river and the deeper inflowing saline countercurrent. Although some direct conversion of Mn_p to Mn_d may occur, the observed Mn_d distributions argue against extensive conversion of Mn_p to Mn_d within the water column. It should be noted that all of these surveys were conducted at times of high riverine Mn concentrations — different behaviours may occur under different conditions.

A single study of dissolved lithium in the St. Lawrence Estuary and Gulf of St. Lawrence (Stoffyn-Egli 1982) shows conservative mixing with little scatter and a very close correlation between lithium and salinity. The river concentration was 2.0 $\mu g \cdot L^{-1}$ (August 1979) increasing to 180 $\mu g \cdot L^{-1}$ at a salinity of 34.8. A study of the estuarine mixing of boron (Pelletier and Lebel 1978) also shows conservative behavior. Dissolved chromium concentrations measured on samples from the same cruise as those for Li showed extensive and rapid removal in the uppermost regions of the Estuary (Campbell and Yeats 1984). The concentration decreased from 0.7 $\mu g \cdot L^{-1}$ in the river to 0.4 $\mu g \cdot L^{-1}$ at a salinity of 3. Thereafter, between salinities of 3 and 35, the mixing was conservative. The rapid decrease in Cr_d concentrations coincided both with the start of estuarine mixing and with the highest SPM concentrations of the turbidity maximum (see Chapter II). A mercury transect for the Upper Estuary (June 1980, Gobeil et al. 1983) showed low and constant dissolved Hg (generally < 3 ng$\cdot L^{-1}$ with a few results as high as 6 ng$\cdot L^{-1}$) throughout the Estuary. Total mercury concentrations corresponded, like those of total Fe, Mn and Co (Bewers and Yeats 1978), to the SPM distribution, particularly to the high SPM in the turbidity maximum. In the turbidity maximum, total Hg concentrations are as high as 75 ng$\cdot L^{-1}$ while in the river and in the higher salinity water downstream they are generally less than 8 ng$\cdot L^{-1}$. A study of selenium (Takayanagi and Cossa 1985) showed rapid removal of dissolved Se(IV) from 120 ng$\cdot L^{-1}$ in the river to 95 ng$\cdot L^{-1}$ at a salinity of 1 followed by conservative mixing to a final concentration of 20 ng$\cdot L^{-1}$ at 30. Total dissolved selenium showed conservative mixing of river water at 170 ng$\cdot L^{-1}$ with seawater at 50 ng$\cdot L^{-1}$. Dissolved organic selenium concentrations were constant at ≈ 30 ng$\cdot L^{-1}$.

Serodes and Roy (1983) found three distinct patterns for radionuclide distributions in the St. Lawrence Estuary. Their results are total concentrations based on the analysis of flocculated surface water samples collected at five stations between Quebec City and Rimouski. The isotopes ^{137}Cs, ^{144}Ce, ^{226}Ra, and ^{228}Th exhibit maximum concentrations in the most turbid upstream part of the turbidity maximum indicating that the distributions are dominated by the particulate phase. The isotopes ^{7}Be and ^{106}Ru show continuous decreases from maximum concentrations in the river unaffected by the turbidity maximum zone. The atmospherically-derived isotope ^{7}Be shows a more rapid decrease at low salinity than does ^{106}Ru, which may indicate some preferential removal of ^{7}Be in this region. The riverine concentrations of ^{235}U are approximately equal to those in the surface waters of the Lower Estuary with elevated concentrations found in the downstream part of the turbidity maximum. Release of ^{235}U from estuarine sediments would be implied from this distribution. A more restricted second sampling, in which particulate samples were separated from the total samples by centrifugation, showed that Cs, Ce, Ra, and Th are dominated by the particulate phase as expected from the first transect. Particulate fractions are generally less important for Be, Nb, Ru, and U.

Lower Estuary

Distributions of total Mn, Fe, Co, Ni, Cu, Zn, and Cd and dissolved Mn and Fe in the Lower St. Lawrence Estuary (May 1974) were discussed by Bewers and Yeats (1979). They described the vertical profiles in the Lower Estuary for all these metals and compared the average concentrations in surface, deep and bottom waters. All metals show profiles with high concentrations in the surface layer, lower concentrations in the deep water (50-250 m) and increases in the bottom water. On a statistical basis, all the differences between concentrations in the surface and deep layers are significant but deep to bottom water increases for Fe_d, Co_t, Ni_t, and Cu_t are not. The Mn_d increases in the bottom water are by far the most striking. Descriptions of the Upper Estuary metal distributions were extended to the Lower Estuary, both in terms of metal/salinity relationships and metal flux estimates. Removal of Fe_d continued throughout the Lower Estuary. Additional data for Mn_d helped delineate the mixing curve. Although the scatter is quite large, combined Upper and Lower Estuary results suggest a non-linear relationship with augmentation of Mn_d throughout much of the salinity range. The suggestion, from the Upper Estuary results, that Ni follows a similar relationship to Mn is strengthened somewhat by the addition of Lower Estuary results. In general, estuarine processes evident from the Upper Estuary results continue through the surface and intermediate depth waters of the Lower Estuary. A new feature is the observation of elevated concentrations in the bottom waters (> 250 m depth) of the Lower Estuary.

Subramanian and d'Anglejan (1976) also observed continuation in the Lower Estuary of Upper Estuary metal/salinity relationships for Mn and Fe. Stoffyn-Egli (1982) observed continuation of the conservative lithium/salinity relationship throughout the Lower Estuary and Gulf of St. Lawrence. Similarly, Campbell and Yeats (1984) observed the general continuation of a conservative chromium/salinity relationship for salinities > 5 but did obtain some anomalously low results in the Pte. des Monts section. No explanation was given for these observations.

Some resampling for trace metals in the waters of the Lower St. Lawrence Estuary was conducted in August 1979 (BIO cruise 79-024) using the improved sampling and analytical techniques that had been developed in the intervening years. Sampling was conducted with specially modified and tested Go-Flo samplers (Bewers and Windom 1982) and filtration with well-tested equipment and methods. Although the analytical methods had not changed in principle (APDC/MIBK extraction/flameless AA), many small improvements had been made since 1974. The results of these improvements can be seen in the improved ability to measure deep-sea metal concentrations (Yeats and Campbell 1983).

Dissolved metal results from three stations in the Lower Estuary (cruise 79-024, August, 1979), are shown in Fig. VI.1. These samples were collected to investigate further the distributions of trace metals in the bottom waters (Fig. VI.1). The results from cruise 74-006 clearly showed elevated near bottom manganese concentrations that resulted from release of dissolved manganese from the sediments (Yeats et al. 1979) and similar

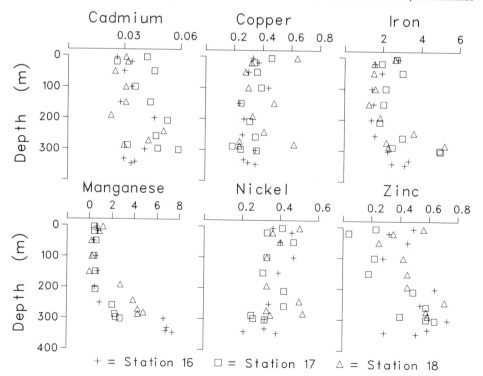

FIG. VI.1. Profiles of dissolved Cd, Cu, Fe, Mn, Ni, and Zn at three stations in the Lower St. Lawrence Estuary. All concentrations are in μg/L.

but less dramatic increases in several other metals (Bewers and Yeats 1979). It was not clear whether the deep water increases in Fe, Zn, or Cd were due to similar sediment releases or some other processes. One of the aims of this resurvey of the Lower Estuary was to investigate this problem.

The new Lower Estuary results show generally similar features to those observed for Mn_d, Fe_d, Co_t, Ni_t, Cu_t, Zn_t, and Cd_t in 1974 (Bewers and Yeats 1979) but the levels are all lower than in the 1974 results. Three possible reasons for the decreased levels are: filtered samples instead of unfiltered ones (except for Mn and Fe), fall sampling instead of spring, and improved methods of collection and analysis. All three probably contribute to the observed differences. Increasing concentration with depth are seen for dissolved Mn, Fe, Zn, and Cd and uniform concentrations for Ni and Cu. The manganese profiles are very similar to our previous observations except that the magnitudes of the near-bottom increases are somewhat smaller. The iron distribution also shows near-bottom increases very much like those of manganese. Near-bottom Fe concentrations are significantly ($P < 0.01$) greater than those in the deep water. This result is surprising since the fast kinetics of iron II oxidation should rapidly remove any Fe^{II} released from the sediments to the particulate phase. It is conceivable that the measured concentrations largely represent colloidal or very fine particulate Fe^{III} that passed through the 0.4 μm filters. Zinc concentrations also increase towards the bottom. There is more scatter in the results but the difference between deep and bottom waters is still significant at $P < 0.05$. It is quite possible that regeneration of biogenic debris in the deep water, which is evident from increased silicate concentrations, is also responsible for the observed Zn increases. Cadmium also shows concentration increases with depth, but in this case the maximum concentration is within the deep water, not in the near-bottom water. Nutrient-related processes are probably responsible for these observations although releases from sediments could make some contribution to the increases in Cd.

An extensive program to study the diagenesis of trace metals in the sediments of the Lower Estuary and Gulf and their fluxes across the benthic boundary layer has been conducted by Sundby, Silverberg and coworkers (Sundby et al. 1981; Sundby and Silverberg 1985; Gendron et al. 1986; Gobeil et al. 1987). These studies focussed on measurements of dissolved and particulate metals in sediment cores, but have obvious implications for investigations of trace metal behaviour in the water column.

Box core samples from sediments underlying the deep waters of the Lower Estuary and Gulf have been analysed for Fe, Mn, Co, and Cd in the solid phase (both for total metal and the hydroxylamine/acetic acid extractable fraction) and for dissolved Fe, Mn, and Cd in the pore water. The concentrations of extractable Fe, Mn, and Co decrease from maximum concentrations in the oxidized, surface sediments to uniformly low concentrations in the reduced, subsurface sediments. Pore water concentrations of Mn and Fe increase from very low levels at the sediment surface to maximum concentrations within the reduced layer. Elevated Mn concentrations penetrate further up into the oxidized layer than do dissolved Fe concentrations, which would imply that Fe fluxes from the sediments to the water column are less important than those of Mn. Cadmium shows an entirely different distribution. The concentrations of the hydroxylamine/acetic acid extractable fraction increase with depth through the surface layer, then remain reasonably constant in the subsurface layer. The dissolved Cd results are more erratic but tend to be higher in the surface layer than in the upper part of the subsurface layer. Deeper within the reduced layer, dissolved Cd concentrations increase with depth.

Although the main purpose of these papers is to describe the sediment diagenesis of these metals and other features of their sediment geochemistry, there are important connections between sediment processes and metal behaviour in the water column. This interaction has been described for manganese (Yeats et al. 1979), and dissolved Mn fluxes from the sediments to the water column are calculated in one of the sediment geochemistry papers (Sundby and Silverberg 1985). The Co results are similar to those for Mn, both in the vertical structure observed in extractable and pore water concentrations within a core, and in the increase in extractable concentrations found in cores further "upstream" (i.e. landward) in the deep Laurentian Channel water. These similarities would suggest a flux of dissolved Co into the water column, and a Co cycle similar to that described for Mn (Yeats et al. 1979; Sundby et al. 1981) would be expected for Co (Gendron et al. 1986). Diagenesis of Cd is obviously quite different from that of Mn, Fe, or Co, but relatively high pore water Cd concentrations in the surface sediments may also result in a flux of dissolved Cd to the overlying water column. As yet, no strong evidence has been obtained for elevated Co or Cd in the deepest waters of the Laurentian Trough.

Open Gulf of St. Lawrence

The only general description of trace metal concentrations in the open Gulf of St. Lawrence is the rather old study of total Fe, Co, Ni, Cu, Zn, Cd, and Pb in the waters of the Gulf by Bewers et al. (1974), using samples collected on BIO cruise 72-017. This study was seriously limited by the dependence on flame atomic absorption spectrophotometry for trace metal analysis but, considering this limitation, gave reasonably reliable average concentrations for total and dissolved Fe and total Ni, Cu, and Zn. Cobalt, cadmium and lead levels were below their detection limits in almost all samples.

The filtered and unfiltered samples were indistinguishable for Ni, Cu, and Zn which is not surprising considering the limitations of flame AA and the low levels of particulate matter in the Gulf. Average concentrations of total Ni, Cu, and Zn were 0.35, 0.46 and 1.52 $\mu g \cdot L^{-1}$, respectively, which were very similar to open ocean concentrations being reported at that time. No geographic or depth variation was seen for these metals. Dissolved iron concentrations were more variable ($0.9-13.8$ $\mu g \cdot L^{-1}$) and showed a non-linear (loss) relationship with salinity. This observation extends the observed estuarine loss curve for Fe_d to higher salinities. Total iron distributions were similar to those of Fe_d but with higher concentrations as expected from the relatively high iron content of both inorganic and biogenic SPM.

Newer results from Cabot Strait (Fig. VI.2) results show more scatter for several of the metals than do the Lower Estuary results from the same cruise (Fig. VI.1). Techniques were identical so the increased scatter would suggest some real but unresolved features of the data. Water flows out of the Gulf predominantly at shallower depths in the southern half of Cabot Strait; the inflow of more saline water occurs at greater depths more or less in the centre of the Strait (see Chapter I). Metal distributions that predominantly reflect the influence of high metal concentrations in the freshwater input should reflect this flow pattern. The concentrations of Mn, Fe, Ni, Cu, and Zn decrease with depth so concentrations in the outflow are greater than in the inflow. These differences are significant for all of the metals except Ni and reflect the importance of the freshwater input in all the distributions. Cadmium concentrations, on the other hand, increase with depth in a manner that is related to the nutrient-like behaviour of cadmium. Cd, Ni, and Zn all show distributions closely related to those of the nutrients in the deep ocean. In the Gulf of St. Lawrence, however, only Cd has a nutrient-like distribution because river concentrations of Cd are relatively small. For Zn and Ni any nutrient related signal is masked by the river input signal.

Saguenay Fjord

Dissolved iron and manganese distributions in the Saguenay Fjord have been described by Yeats and Bewers (1976) and Bewers and Yeats (1979). Samples were collected in May 1974. The dissolved iron versus salinity relationship (Yeats and Bewers 1976) indicates extensive estuarine removal of iron occurring over a broad salinity range. The Saguenay River iron concentration observed during this survey, 702 $\mu \cdot L^{-1}$ (actually at a salinity of 0.5 at the head of the Fjord), was extremely high. The dissolved manganese versus salinity relationship (Bewers and Yeats 1979) showed similar, but less extensive, removal of Mn_d from the surface water. This salinity relationship is unlike that observed in the St. Lawrence Estuary or in most other estuarine studies where

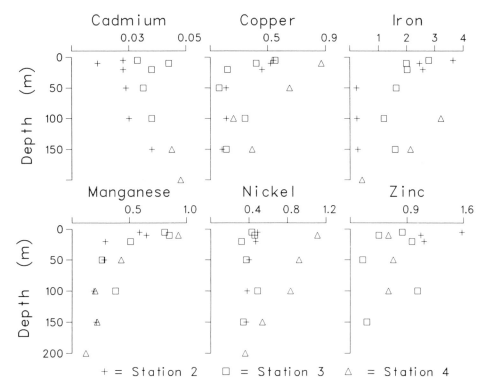

FIG. VI.2. Profiles of dissolved Cd, Cu, Fe, Mn, Ni, and Zn at three stations in Cabot Strait. All concentrations are in μg/L.

conservative relationships or injection of dissolved Mn are generally observed. The general isolation of most of the Saguenay surface water from direct interaction with the bottom sediments (see Chapter I) may be responsible for this difference in behaviour. Elevated bottom water dissolved manganese concentrations resulting from release of Mn_d from the sediments were observed in this study. A conservative dissolved molybdenum versus salinity relationship based on surface samples from the same cruise has been reported by Kulathilake and Chatt (1980).

Total Co, Ni, Cu, Zn, and Cd distributions were described in terms of the average concentrations in the various hydrodynamic regions of the Fjord rather than in terms of the relationships with salinity (Yeats and Bewers 1976). Higher concentrations of all five metals are found in surface waters than in the underlying waters, mostly as a result of input of metals from the Saguenay River. Increased near bottom concentrations were also observed for copper and nickel.

The results of a resurvey of the Saguenay Fjord in May 1976 (cruise 76-006) are listed in Table VI.3. Average dissolved metal concentrations are shown for the surface waters, deep waters of the eastern basin, and intermediate and deep waters of the western basin. The results are generally quite comparable to the dissolved Mn and Fe and total Co, Ni, Cu, Zn, and Cd results of the 1974 survey. In the 1976 data, only dissolved manganese shows a significant concentration increase in the near-bottom water. Manganese shows lower concentrations in the eastern than western basins while Co and Cd show the opposite trend.

Yeats and Bewers (1976) calculated particulate iron concentrations from the difference between total (unfiltered) and dissolved (filtered) Fe measurements. They then computed particulate iron content of the SPM by dividing Fe_p by the SPM concentration. SPM samples from this cruise were collected by pressure filtration directly from the sampling bottles. Subsequent work has shown that this method leads to underestimation

TABLE VI.3. Dissolved metal concentrations in the Saguenay Fjord ($\mu g \cdot L^{-1}$).
Concentrations are given $\pm$ 1σ, with the number of samples in parentheses.

	Surface	Western Basin intermediate	Western Basin deep	Eastern Basin deep
Depth	< 10 m	50-100 m	> 200 m	> 50 m
Salinity	0.4 - 19	25 — 30.3	≈30.8	23 - 30
Al	78 ± 33 (14)	21 ± 11 (17)	13 ± 4 (8)	15 ± 8 (21)
Mn	10.9 ± 6.5 (16)	2.7 ± 0.9 (14)	5.9 ± 2.1 (11)	1.7 ± 0.4 (21)
Fe	52 ± 24 (13)	11 ± 6 (16)	7 ± 3 (11)	12 ± 7 (22)
Co	0.040 ± 0.036 (16)	0.026 ± 0.013 (9)	0.009 ± 0.005 (3)	0.018 ± 0.009 (3)
Ni	0.50 ± 0.12 (17)	0.46 ± 0.08 (15)	0.45 ± 0.09 (7)	0.40 ± 0.07 (15)
Cu	0.84 ± 0.33 (17)	0.40 ± 0.12 (7)	0.50 ± 0.31 (3)	0.68 ± 0.16 (4)
Zn	1.9 ± 0.8 (16)	0.8 ± 0.3 (13)	1.0 ± 0.4 (7)	0.9 ± 0.4 (14)
Cd	0.076 ± 0.054 (17)	0.051 ± 0.025 (12)	0.044 ± 0.015 (8)	0.074 ± 0.019 (10)

of the SPM concentration in turbid waters. As a result, Fe content of the SPM calculated in this way will overestimate the true values, but this does not adversely affect the validity of the general conclusions of the paper.

Sundby and Loring (1978) measured the Fe and Mn content of the SPM (as well as Si, Al, Ca, Mg, and K) by direct measurement of the metal content of SPM samples. They found that iron varied from ≈5% in surface waters to >9% in near bottom waters of the western basin of the Fjord. The distributions in May 1974 and September 1974 were quite similar. In May, manganese showed low concentrations in surface water (< 1000 $\mu g \cdot g^{-1}$), increasing to maximum concentrations of > 7500 $\mu g \cdot g^{-1}$ at 150 m depth in the western basin. In September, the deep water maximum was displaced westward and to shallower depths. Surface concentrations were also higher (> 2000 $\mu g \cdot g^{-1}$).

The particulate iron and aluminum results from cruise 76-006 (Table VI.4) show very similar Fe/Al ratios to those reported by Sundby and Loring (1978). The Mn/Al ratios are somewhat different in that the highest values are observed in the bottom water, not at intermediate depth. The bottom water ratios are very similar to those of Sundby and Loring. The high Mn region at intermediate depth may simply have been missed as the sampling density was rather low. There is no evidence for particles with high Fe or Mn content in the eastern basin of the fjord.

Cossa and Poulet (1978) also observed lower Mn, Zn, Pb, and Cd content of SPM in surface waters of the Fjord than in the deep waters of the western basin. They attributed the metal enrichment in the deep water to authigenic formation of Mn oxides resulting from the precipitation of Mn released from the bottom sediments and subsequent absorption of Zn, Pb, and Cd onto these oxide surfaces. Sundby and Loring (1978), on the other hand, attribute their observations of high Mn in the deep water to advection of particles with high Mn content into the deep water of the Fjord from the St. Lawrence

TABLE VI.4. Particulate Fe and Mn in the Saguenay Fjord. Concentrations are normalized to the particulate Al concentrations (w/w) and are given $\pm$ 1δ, with the number of samples in parentheses.

	Fe/Al	Mn/Al
Western Basin		
surface (0-10 m)	0.52 ± 0.11 (16)	0.0082 ± 0.0030 (15)
intermediate (50-100 m)	0.56 ± 0.15 (9)	0.012 ± 0.006 (5)
deep (> 200 m)	0.87 ± 0.12 (7)	0.018 ± 0.013 (7)
Eastern Basin		
surface (0-10 m)	0.55 ± 0.07 (7)	0.0076 ± 0.0026 (6)
deep (> 50 m)	0.41 ± 0.07 (9)	0.0060 ± 0.0025 (10)

Estuary. Observation of dissolved Mn releases from the fjord sediments and similar *in situ* enrichment of the Mn content of SPM in the Lower St. Lawrence Estuary (Yeats et al. 1979) would suggest that the Mn concentrations in the Saguenay SPM are also internally generated as suggested by Cossa and Poulet (1978). Advection would be responsible for displacement of the maximum. The differences between the observations for May 1974 and May 1976 are also consistent with internal generation, rather than advection, of these features. In May 1974, the dissolved manganese concentrations in the bottom water are very high (Bewers and Yeats 1978) and the particulate manganese maximum is well developed (Sundby and Loring 1978). This picture of the distribution is very similar to that reported by Yeats et al. (1979) for the Lower St. Lawrence Estuary. In May 1976, the maximum dissolved Mn concentrations in the bottom water are lower than in 1974 (Fig. VI.3) and the particulate Mn maximum is located closer to the bottom.

Ellis and Chattopadhyay (1979) have also reported the concentrations of 28 metals in three SPM samples collected from the surface waters of the Fjord. There is good agreement between these results and those reported by Cossa and Poulet (1978) and Sundby and Loring (1978) for the elements Ca, Fe, K, Mn, and Zn. For most of the remaining 23 elements no other data are available.

Water samples collected on cruise 76-006 (May 1976) have also been used to study the trace metal geochemistry in the estuarine mixing zone in more detail. Both dissolved metal concentrations (solvent extraction of filtered water samples) and non-detrital particulate metal concentrations (25 % acetic acid leaches of SPM samples collected on 0.4 μm Nuclepore filters) were measured for samples with salinities from the total range found in the Fjord. The results are illustrated in Fig. VI.3 to VI.5 as plots of dissolved and non-detrital particulate metals versus salinity. The Mn_d/salinity relationship shows agreement with the manganese results from May 1974 (Bewers and Yeats 1979) with better resolution of the behaviour in the 0-3 salinity range. Manganese concentrations increase from 14.1 ± 0.7 $\mu g \cdot L^{-1}$ at a salinity of 0.4 (the lowest salinity water sampled) to a maximum concentration at a salinity of ≈ 2. Five samples with salinities between 1.9 and 2.5 gave Mn_d concentrations of 14.3 to 22.2 $\mu g \cdot L^{-1}$. From this maximum, concentrations decrease in a non-conservative manner indicative of removal of Mn_d to the particulate phase. Non-detrital particulate Mn shows a complimentary salinity relationship. Concentrations decrease from 400 μg $Mn \cdot g^{-1}$ SPM at 0.5 to 200-260 $\mu g \cdot g^{-1}$ in the 2-5 salinity range followed by a gradual increase to 460-500 $\mu g \cdot g^{-1}$ at a salinity of 26. The samples from the eastern basin have lower concentrations than ones of similar salinity from the western basin.

Maxima in dissolved manganese at low salinity have been observed in several estuaries and have generally been attributed to diagenetic release of Mn from estuarine sediments or remobilization from estuarine SPM caused by salinity increases. It might appear that, in our case, the excess Mn_d comes directly from the SPM. Loss of 150 $\mu g \cdot g^{-1}$ of non-detrital Mn from ≈ 10 $mg \cdot L^{-1}$ of SPM (SPM concentration estimated from Al_p) would give 1.5 $\mu g \cdot L^{-1}$ of additional Mn_d. This release of Mn_d will account for much, if not all, of the observed increase. However, SPM, as calculated from Al_p, also increases in this region so a source of SPM is needed. This source is almost certainly resuspension of underlying sediments (the westernmost station was 40 m deep) or erosion from adjacent sides of the basin. Resuspension of SPM with lower non-detrital Mn content could then account for the decrease in non-detrital Mn between salinities of 0.4 and 5. Remobilization of Mn_d from these same sediments could also be a source for some, or all, of the increased Mn_d observed in the region. Temporal variability, with observed concentrations at downstream stations simply reflecting river concentrations from previous days, could also account for some of these observations.

The other metals show distributions of dissolved (Fig. VI.4) and non-detrital particulate (Fig. VI.5) phases that are generally similar to those of manganese. Poorer precision and less extreme concentrations mean that these features are not as clearly illustrated as was the case with manganese. Surface dissolved Al and Fe concentrations on five stations near the head of the Fjord were extremely high (> 200 $\mu g \cdot L^{-1}$) and have not

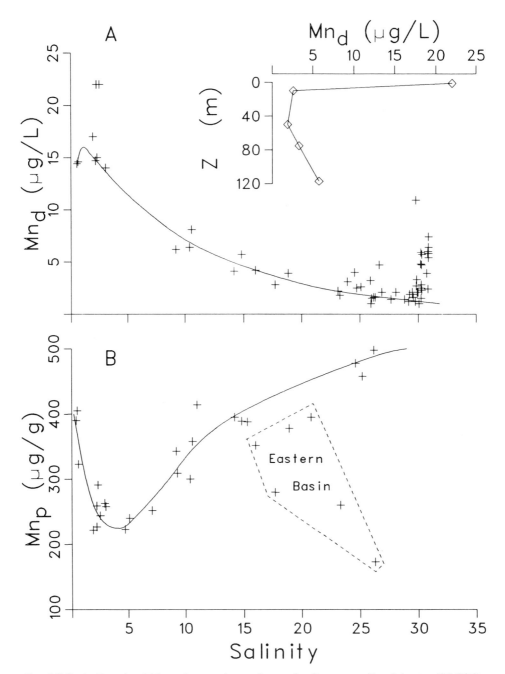

FIG. VI.3. A. Dissolved Mn-salinity relationship in the Saguenay Fjord (cruise 76-006). Inset shows a typical Mn_d depth profile for a station close to the head of the Fjord; B. Non-detrital particulate Mn (expressed relative to the total suspended particulate matter) — salinity relationship in the Saguenay Fjord.

been plotted. The reason for these observations is a puzzle. One possible explanation is that the filtration failed and the "filtered" samples contained some unfiltered water. However, this does not seem to be the case as none of these samples were visibly turbid immediately after filtration and Al_p and Fe_p results for these samples are not correspondingly low. It would seem more probable that the high concentrations represent very

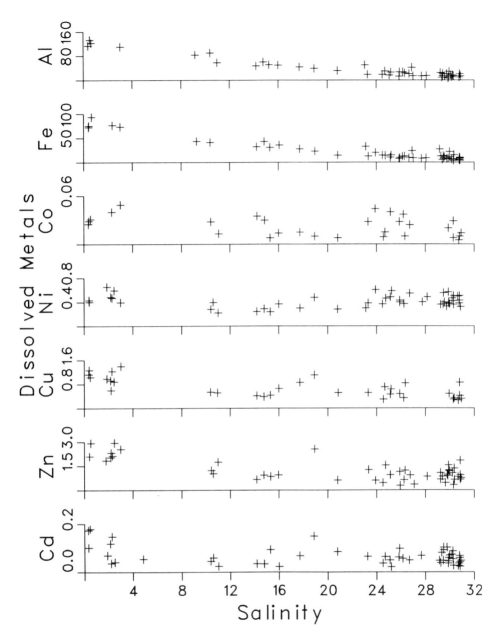

FIG. VI.4. Dissolved Al, Fe, Co, Ni, Cu, Zn and Cd–salinity relationships in the Saguenay Fjord. Concentrations are in μg/L.

fine particles or colloids that were forced through the 0.4 μm filters. The two dissolved
Mn results at 22 μg·L^{-1} are from samples with high Al and Fe but the other three high
Al and Fe samples show little or no elevation in Mn content.

The distributions of dissolved Co and Ni look most like that of manganese. Both
of these metals have maximum concentrations at salinities of $\approx$2, and non-linear
decreases for the rest of the surface layer. The other metals, Al, Fe, Cu, Zn, and Cd,
show maximum concentrations in the river water and nonlinear (loss) decreases throughout the mixing zone. For Cu and Cd, the eastern basin waters have noticeably higher
concentrations. For Co the eastern basin intermediate depth water has high

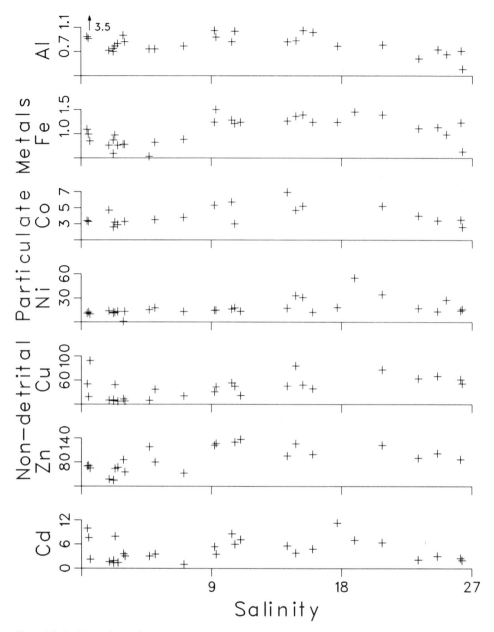

FIG. VI.5. Non-detrital particulate (expressed relative to the total SPM) Al, Fe, Co, Ni, Cu, Zn, and Cd-salinity relationships in the Saguenay Fjord. Al and Fe concentrations are expressed in %, all others in μg/g SPM.

concentrations. These anomalous concentrations can also be identified in Table VI.3.

None of the other non-detrital particulate metal distributions (Fig. VI.5) exactly mimic the non-detrital manganese distributions (Fig. VI.3B) although all but the Ni distribution have minimum concentrations in the 2–5 salinity range. Most of the distributions, however, have maximum concentrations at mid-salinity, and none show the anomalous concentrations for the eastern basin waters that were seen for manganese.

If the Saguenay surface water can be described by a simple model of mixing of freshwater and seawater without interaction with the sediments (at least for salinities > 2) then the losses of dissolved metals within the mixing zone should correspond to the gains in non-detrital particulate metal. This is indeed the case. If allowance is made for the settling of SPM out of the surface layer, the increases in non-detrital particulate metal concentrations can account for 35-144 % of the losses of dissolved metals.

Manganese Geochemistry

The marine geochemistry of manganese is perhaps more interesting than that of any of the other trace metals. The range of concentrations observed in the marine environment is quite large, decreasing from ≈ 10 $\mu g{\cdot}L^{-1}$ in rivers to only 20 $ng{\cdot}L^{-1}$ in the deep North Atlantic. There also exists a broad range of chemical reactions involving manganese in the ocean. These include adsorption/desorption reactions with ambient particles and reactions involving the oxidation of Mn^{II} to insoluble MnO_2. Manganese dioxide can be a host for the adsorption of other trace metals. Dissolved manganese can also be removed by incorporation into biological material. As a result of this diversity of processes, Mn has attracted more scientific attention than any of the other trace metals.

Estuarine distributions of Mn have generally shown addition of Mn_d in the low salinity range, often with removal at higher salinities (Evans et al. 1977). As a result, the Mn_d versus salinity relationship is often complex showing overlapping concave and convex curves. Since both the underlying sediments and SPM can be sources of Mn_d, considerable scatter is often seen in the Mn_d versus salinity plots. Laboratory simulations of estuarine mixing have generally shown removal of Mn along with most other metals during mixing (Sholkovitz 1976).

In the St. Lawrence Estuary, the Mn versus salinity relationship for May 1974 indicates slight addition of Mn_d in the mixing zone (Bewers and Yeats 1979). Rapid decrease of Mn_p concentrations from riverborne SPM to estuarine SPM (Cossa and Poulet 1978; Gobeil et al. 1981) would suggest that some, or all, of the additional Mn_d could come from the suspended particulate phase. The reduction in Mn_p, however, is very rapid and would be expected to lead to a large increase in Mn_d over a narrow salinity range rather than the smaller increases observed throughout much of the mixing zone. This would suggest that release of Mn_d from the sediments and augmentation of turbidity zone SPM with resuspended sediment low in Mn_p have major effects on both Mn_d and Mn_p distributions. A sedimentary source for the excess Mn_d makes good chemical sense since the general mode of removal of Mn^{2+} is oxidation to insoluble MnO_2 rather than simple adsorption of manganese to suspended particulate surfaces. In oxygenated waters the oxidation will not be a reversible reaction. However, in the sediments, reduction of MnO_2 under anoxic conditions is a well-established reaction that results in the formation of Mn^{2+} and its diffusion into surficial oxidized layers and/or the overlying water. On the other hand two processes can contribute to direct release of Mn from estuarine SPM. They are desorption of Mn^{2+} from particles as a result of increased ionic strength and disturbance of the Mn^{2+} - MnO_2 equilibrium by the decrease in O_2 content that is frequently observed in estuaries.

In the Saguenay Fjord, dissolution of Mn from Mn_p or remobilization of Mn_d from fjord sediments results in additions of Mn_d at very low salinity, while at higher salinity, removal of Mn_d from solution is observed. The Fjord surface waters, having salinities of > 2 are sufficiently isolated from the sediments that releases from sediments cannot have a direct effect on the surface distributions. Chemical precipitation reactions can thus be observed without the complications introduced by sediment remobilization.

In the deep water of the Lower St. Lawrence Estuary, unequivocal evidence for remobilization of Mn_d from estuarine sediments and subsequent precipitation of Mn_d within the water column has been obtained (Yeats et al. 1979). Deep water SPM was first shown to have very high Mn content by Sundby (1977). The maximum concentrations were observed 40-100 m above the bottom rather than in the very deepest samples. The Mn content of these particles was shown to be 4-5 times the maximum concentrations in the underlying sediments. It was therefore evident that simple resuspension of bottom sediments could not explain the observations. Either preferential resuspension of fine Mn-enriched particles or precipitation of Mn_d on SPM were suggested as possible mechanisms. The potential importance of this enrichment process as a source of Mn to the pelagic ocean was identified in this study.

A more extensive follow-up paper (Yeats et al. 1979), incorporating dissolved and particulate metal results from the Lower Estuary, proved conclusively that Mn is transferred from the sediments to the deep-water particles via the dissolved phase. The main findings of this study are illustrated in Fig. VI.6. Dissolved Mn decreases very rapidly from high levels in the deepest samples to very low levels 60 m off the bottom. Mn_d concentrations are then constant from this depth to the lower salinity surface mixed layer where slightly higher concentrations are observed as a result of the estuarine freshwater input. Particulate Mn concentrations are high in the deepest samples and increase to maximum concentrations 30-100 m above the bottom. Above this depth, Mn_p concentrations decrease to the surface. SPM concentration is highest at the surface, decreasing to uniformly low levels throughout most of the water column with an approx-

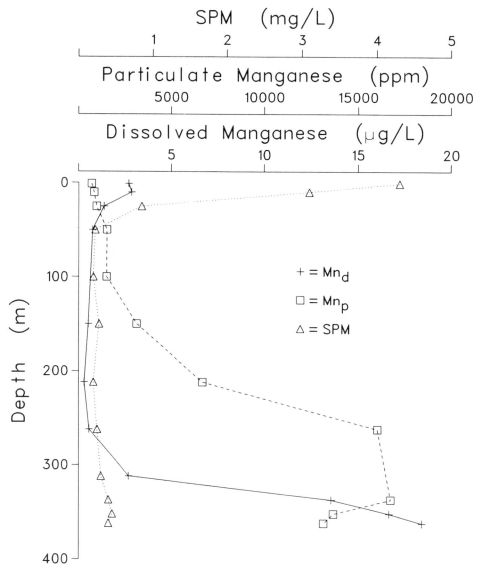

FIG. VI.6. Dissolved Mn, particulate Mn and total suspended matter profiles at a station in the Lower St. Lawrence Estuary (from Yeats et al. 1979.)

imately two-fold increase in the bottom 50 m. Release of Mn_d from the bottom sediments and precipitation onto SPM in the deep waters are responsible for these observations.

The maximum in the Mn content of the SPM, 30-100 m above the bottom, is an important feature of the distributions that is consistent with the enrichment processes described here. Two features of the bottom layer SPM distribution are the decreasing contribution of the resuspended sediment to the SPM and the increasing importance of small-sized particles with distance above the bottom. Preferential association of precipitating Mn_d with the smallest particles, combined with decreased dilution with resuspended SPM, will result in a maximum in Mn_p some distance above the sediment surface.

This behaviour of manganese in the Lower Estuary results in an internal cycling of riverborne dissolved and particulate Mn mediated by remobilization processes in the sediments. It results in the transformation of estuarine particulate Mn, which settles out in the Lower Estuary, to dissolved and fine-grained particulate forms that have sufficiently slow settling rates to remain in suspension and be transported offshore. Sundby et al. (1981) discuss a flow model that describes these processes in detail. They also present electron microscope data that confirm the existence of very fine particles that are very highly enriched in manganese. This model describes internal recycling of Mn between sediments and water and between dissolved and particulate phases with the net result that Mn is concentrated, perhaps after a number of cycles, in fine-grained particles some of which escape the system in the seaward flowing surface waters. The trace metal flux model described by Bewers and Yeats (1977) illustrates just such an accumulation of Mn_p on the particles exported at Cabot Strait.

Gulf Models and Budgets

We have made several attempts over the years to relate measurements of metal distributions in the Gulf of St. Lawrence and St. Lawrence Estuary to more general questions of fluxes of metals through estuarine and coastal environments and anthropogenic effects of metal levels and fluxes. None of the metal flux estimates in these models will be highly precise or accurate. The models will, at best, give a semi-quantitative estimate of the fluxes as well as contribute to our understanding of the metal behavior.

Our first try at a trace metal model for the Gulf of St. Lawrence (Bewers and Yeats 1977) was part of an attempt to estimate the net influxes of dissolved and particulate metals to the coastal zone accounting for processes occurring in estuaries. In order to do this, a multi-box model for the Gulf was constructed, and metal and SPM data collected at the box boundaries between 1972 and 1976 were used to calculate fluxes at the various boundaries. Salt and mass conservation were used to estimate water fluxes. From this model the proportions of the dissolved and particulate metal inputs to the Gulf that are removed to Gulf sediments could be estimated. It was found that > 93 % of the SPM input, 93 % of Fe_p, 90 % of Mn_p, 54 % of Fe_d, and lesser amounts of Co_t, Ni_t, Cu_t, Zn_t, and Cd_t were removed. Dissolved Mn was generated either from particulates or the sediments. Concentrations in sedimenting material, calculated from these removal rates, were in reasonably good agreement with measurements of metal contents of Gulf sediments. The calculated proportions of metals removed from the Gulf were then used to account for estuarine and coastal zone removal of metals in an attempt to improve estimates of global metal residence times in the ocean. The residence times calculated from this model have generally been confirmed by other more recent estimates.

This trace metal model has been refined by incorporating more recent metal results and some estimates of the internal biological cycling in the Gulf (Bewers and Yeats 1983). The resulting box model, segmented in both vertical and horizontal direction, and calculated separately for three different seasons, is predominantly a nutrient model since internal biological cycling was included. Metals have been included although the density of sampling for the metals was not as good as for the nutrients. On the basis of this model, two thirds of the annual input of SPM to the Gulf is lost to the sediments in the Gulf. The loss of inorganic particulate matter within the system amounts to 86 % of the inputs but there is a net gain of organic particulate material that results in a lower overall loss of suspended matter. The loss of inorganic material corresponds closely to the loss of total iron of 84 %. Chapter X considers the extensions of this model possible in view of the other models discussed in this book.

These models can serve to summarize the findings on metal behaviour in the Gulf. Figure VI.7 illustrates these results by showing the net outflow of metals as a percent of the freshwater input at various sections in the Estuary and Gulf models. The first section at Pte. aux Orignaux is in the middle of the turbidity maximum zone, the second at Ile Rouge is at the downstream boundary of the Upper Estuary, the third at Pte. des Monts is at the boundary between the Lower Estuary and Gulf and the fourth at Cabot Strait is at the eastern boundary of the Gulf of St. Lawrence. This picture is based on spring runoff conditions in the river. The SPM flux increases markedly between the river and Pte. aux Orignaux then drops to only a few percent of the river input at Cabot Strait. At Ile Rouge the outflow is equal to the river input. The Fe_p flux at Pte. aux Orignaux is slightly greater than the river flux but much less elevated than is the SPM flux. Mn_p and Co_p fluxes at Pte. aux Orignaux are both less than the river flux. These estuarine fluxes are indicative of processes that result in the preferential removal of particulate metal compared to SPM. Throughout the rest of the system, particulate metal effluxes decrease such that by Cabot Strait Mn_p and Fe_p fluxes are < 3 % of their respective inputs. Dissolved iron and total (mostly dissolved) zinc fluxes are significantly reduced (by ≈ 50 %) in the Upper Estuary, total Cu and Cd fluxes

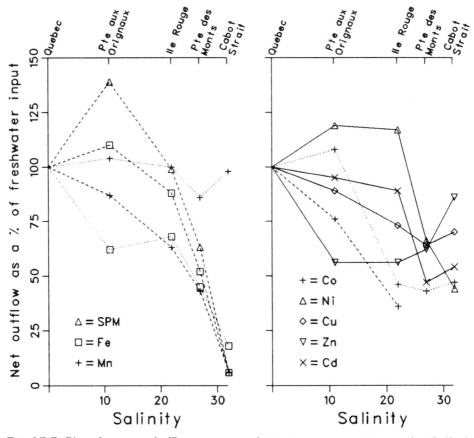

FIG. VI.7. Plot of net metal effluxes versus salinity at various sections in the Gulf of St. Lawrence. Dashed lines = particulate phase; dotted lines = dissolved phase; solid lines = total metal (redrawn from Yeats and Bewers 1983).

are slightly reduced and total Ni and dissolved Mn and Co fluxes actually increase slightly. The estuarine reduction in Fe_d flux continues throughout the rest of the system such that the final net flux at Cabot Strait is only 20 % of the freshwater input. Final Co, Ni, Cu, Zn, and Cd effluxes are 50-75 % of the freshwater inputs. Dissolved Mn is unusual in that the final net efflux at Cabot Strait is equal to the freshwater input indicating no net removal of dissolved Mn in the Gulf of St. Lawrence.

References

BEWERS, J. M., AND H. L. WINDOM. 1982. Comparison of sampling devices for trace metal determinations in seawater. Mar. Chem. 11: 71-86.

BEWERS, J. M., AND P. A. YEATS. 1977. Oceanic residence times of trace metals. Nature (Lond.) 268: 595-598.

1978. Trace metals in the waters of a partially mixed estuary. Estuarine Coastal Mar. Sci. 7: 147-162.

1979. The behavior of trace metals in estuaries of the St. Lawrence basin. Nat. Can. (Que.) 106: 149-161.

1983. Transport of metals through the coastal zone, p. 146-163. In J.B. Pearce [ed.] Reviews of water quality and transport of materials in coastal and estuarine waters. Int. Coun. Explor. Sea Cooperative Res. Rep. 118, Copenhagen.

BEWERS, J. M., I. D. MACAULEY, AND B. SUNDBY. 1974. Trace metals in the waters of the Gulf of St. Lawrence. Can. J. Earth Sci. 11: 939-950.

BOYLE, E. A., S. S. HUESTED, AND B. GRANT. 1982. The chemical mass balance of the Amazon plume-II: copper, nickel and cadmium. Deep-Sea Res. 29: 1355-1364.

CAMPBELL, J. A., AND P. A. YEATS. 1984. Dissolved chromium in the St. Lawrence Estuary. Estuarine Coastal Shelf Sci. 19: 513-522.

COSSA, D., AND S. A. POULET. 1978. Survey of trace metal contents of suspended matter in the St. Lawrence Estuary and Saguenay Fjord. J. Fish. Res. Board Can. 35: 338-345.

D'ANGLEJAN, B. F., AND R. G. INGRAM. 1976. Time-depth variations in tidal flux of suspended matter in the St. Lawrence estuary. Estuarine Coastal Mar. Sci. 4: 401-416.

DANIELSSON, L.-G., B. MAGNUSSON, S. WESTERLUND, AND K. ZHANG. 1983. Trace metals in the Gota River estuary. Estuarine Coastal Shelf Sci. 17: 73-85.

DUINKER, J.C., AND R.F. NOLTING. 1978. Mixing, removal and mobilization of trace metals in the Rhine estuary. Neth. J. Sea Res. 12: 205-223.

ELLIS, K.M., AND A. CHATTOPADHYAY. 1979. Multielement determination in estuarine suspended particulate matter by instrumental neutron activation analysis. Anal. Chem. 51: 942-947.

EVANS, D.W., N.H. CUTSHALL, F.A. CROSS, AND D.A. WOLFE. 1977. Manganese cycling in the Newport River estuary, North Carolina. Estuarine Coastal Mar. Sci. 5: 71-80.

GENDRON, A., N. SILVERBERG, B. SUNDBY, AND J. LEBEL. 1986. Early diagenesis of cadmium and cobalt in sediments of the Laurentian Trough. Geochim. Cosmochim. Acta 50: 741-747.

GOBEIL, C., D. COSSA, AND J. PIUZE. 1983. Distribution des concentrations en mercure dans les eaux de l'estuaire moyen du Saint-Laurent. Can. Tech. Rep. Hydrogr. Ocean Sci. 17: iii + 14 p.

GOBEIL, C., N. SILVERBERG, B. SUNDBY, AND D. COSSA. 1987. Cadmium diagenesis in Laurentian Trough sediments. Geochim. Cosmochim. Acta 51: 589-596.

GOBEIL, C., B. SUNDBY, AND N. SILVERBERG. 1981. Factors influencing particulate matter geochemistry in the St. Lawrence estuary turbidity maximum. Mar. Chem. 10: 123-140.

KLINKHAMMER, G. P., AND M. L. BENDER. 1981. Trace metal distributions in the Hudson River estuary. Estuarine Coastal Shelf Sci. 12: 629-643.

KULATHILAKE, A. I., AND A. CHATT. 1980. Determination of molybdenum in sea and estuarine water with B-Napthoin oxime and neutron activation. Anal. Chem. 52: 828-833.

PELLETIER, E., AND J. LEBEL. 1978. Determination du bore inorganique dans l'estuaire du Saint-Laurent. Can. J. Earth Sci. 15: 618-625.

SERODES, J. B., AND J.-C. ROY. 1983. Distribution of some radionuclides in the St. Lawrence Estuary, Quebec, Canada. Oceanol. Acta 6: 185-192.

SHILLER, A. M., AND E. BOYLE. 1985. Dissolved zinc in rivers. Nature (Lond.) 317: 49-52.

SHOLKOVITZ, E. R. 1976. Flocculation of dissolved organic and inorganic matter during the mixing of river water and seawater. Geochim. Cosmochim. Acta 40: 831-845.

STOFFYN-EGLI, P. 1982. Conservative behavior of dissolved lithium in estuarine waters. Estuarine Coastal Shelf Sci. 14: 577-587.

SUBRAMANIAN, V., AND B. D'ANGLEJAN. 1976. Water chemistry of the St. Lawrence estuary. J. Hydro. 29: 341-354.

SUNDBY, B. 1977. Manganese-rich particulate matter in a coastal marine environment. Nature (Lond.) 270: 417-419.

SUNDBY, B., AND D. H. LORING. 1978. Geochemistry of suspended particulate matter in the Saguenay Fjord. Can. J. Earth Sci. 15: 1002-1011.

SUNDBY, B., AND N. SILVERBERG. 1985. Manganese fluxes in the benthic boundary layer. Limnol. Oceanogr. 30: 372-381.

SUNDBY, B., N. SILVERBERG, AND R. CHESSELET. 1981. Pathways of manganese in an open estuarine system. Geochim. Cosmochim. Acta 45: 293-307.

TAKAYANAGI, K., AND D. COSSA. 1985. Speciation of dissolved selenium in the Upper St. Lawrence Estuary, p. 275-284. In A.C. Sigleo and A. Hattori [ed.] Marine and estuarine geochemistry. Lewis Publishers, Chelsea, MI.

YEATS, P. A., AND J. M. BEWERS. 1976. Trace metals in the waters of the Saguenay Fjord. Can. J. Earth Sci. 13: 1319-1327.

1982. Discharge of metals from the St. Lawrence River. Can. J. Earth Sci. 19: 982-992.

1983. Potential anthropogenic influences on trace metal distributions in the North Atlantic. Can. J. Fish. Aquat. Sci. 40(Suppl. 2): 124-131.

YEATS, P. A., AND J. A. CAMPBELL. 1983. Nickel, copper, cadmium and zinc in the northwest Atlantic Ocean. Mar. Chem. 12: 43-58.

YEATS, P. A., B. SUNDBY, AND J. M. BEWERS. 1979. Manganese recycling in coastal waters. Mar. Chem. 8: 43-55.

Chapter VII

Trace Metal Geochemistry of Gulf of St. Lawrence Sediments

D. H. Loring

*Marine Chemistry Division, Physical and Chemical Sciences Branch,
Department of Fisheries and Oceans, Bedford Institute of Oceanography,
P.O. Box 1006, Dartmouth, N.S. B2Y 4A2*

Introduction

The Gulf of St. Lawrence sediments are the largest repository for, and source of, transition and heavy metals in the eastern Canadian coastal environment. The metals residing in the sediments have been introduced into the estuarine and coastal environment of the Gulf in solution and as part of, or in association with, solid organic and inorganic particles. This material has been, and is being, supplied from natural and, for some elements, anthropogenic sources and deposited in response to past and present depositional conditions.

This chapter describes the abundance, distribution and partition of zinc (Zn), copper (Cu), lead (Pb), cadmium (Cd), cobalt (Co), nickel (Ni), chromium (Cr), vanadium (V), mercury (Hg), beryllium (Be), lithium (Li), and arsenic (As) in the sediments of the Estuary and the Gulf of St. Lawrence. It also discusses the factors that control the dispersal of these elements, and the natural and anthropogenic sources of the metals in the sediments. It is based on a series of environmental geochemical studies which determined the levels, behaviour, and dynamics of heavy metals in sediments from the open Gulf of St. Lawrence, the St. Lawrence Estuary, and the Saguenay Fjord (Loring 1975; 1976a, b; 1978; 1979; 1981; Loring and Bewers 1978; Smith and Loring 1981; Loring et al. 1983). The reader is referred to Chapter I for a very brief discussion on the morphology and sediments of the Gulf of St. Lawrence. A much more thorough discussion on these topics may be found in Loring and Nota (1973).

Field and Laboratory Methods

Surface sediment samples (219) were collected from different parts of the region with a Van Veen (0.1 m^2) grab sampler (Fig. VII.1). Selected core samples were also obtained from the Estuary and Saguenay Fjord with a 75 mm diameter gravity corer. Representative portions of the grab samples were air dried and stored in air-tight bottles until used for sedimentological, chemical and mineralogical analyses (Loring and Rantala 1977). A preweighed dried sample was wet-sieved to determine the amount of sand (2000-53 μm) and mud ($< 53 \mu$m) size material in each sediment sample. In addition, the size fractions 2000-500 μm, 500-53 μm, 53-37 μm, 37-16 μm, 16-2 μm, and $< 2 \mu$m were removed from 18 samples from different parts of the Gulf system to determine the regional granulometric variations of trace metals. After the removal of sea salts, the major elements (Si, Al, Ti, Na, K, Fe, Mg, Ca, and Mn) were determined using the atomic absorption techniques described by Rantala and Loring (1975). Total trace element concentrations (Zn, Cu, Pb, Co, Ni, Cr, V, Li, Be), and acetic acid (25 % v/v) soluble ("non-detrital"), and insoluble ("detrital"), Zn, Cu, Pb, Co, Ni, Cr, V, and Cd were determined from a dried sample by the atomic absorption techniques described by Loring and Rantala (1977). Arsenic was determined with a colorimetric technique in which samples were digested in a HNO$_3$-HClO$_4$ mixture for 3 hours. Cadmium was determined using graphite furnace atomic absorption tech-

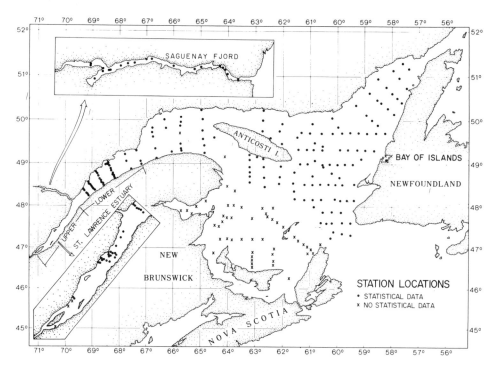

FIG. VII.1.Location of samples used for geochemical analyses. Dots indicate samples for which complete statistical data is available (from Loring 1978).

niques. Mercury was determined by a cold vapour atomic absorption method comparable to that described by Hatch and Ott (1968). The precision for major and minor elements was less than or equal to 5 % and for trace elements generally < 10 %. The relative accuracies for Be, Co, Cr, Cu, Li, Pb, V, and Zn were within ± 13 % of those reported by Flanagan (1973) for the United States Geological Survey marine sediment reference material MAG-1. Cd results were within ± 5 % of those reported for USGS rock standards and Hg results within ± 10 % of the values reported for the USGS rock standards G-1 and W-1. Identification and semi-quantitative analyses of the dominant minerals in the size fractions < 37 μm and for the size fractions 500-53 μm in selected samples were performed by microscopic, X-ray diffraction, and microprobe techniques and reported by Loring and Nota (1973) and Loring (1976b).

Correlation matrices and r-mode factor analyses (Cameron 1968, 1969; Loring 1976a, 1978) have been used to clarify the relationships between the distribution of the elements and the textural and chemical characteristics of the sediments in the different parts of the region. Statistical differences between non-detrital and detrital element concentrations in the different sediment types and between texturally equivalent sediments in each region have been assessed with Student's t-test at the 95 % confidence level.

Abundance and Regional Distribution

The abundance and regional distribution of trace metals in estuarine and coastal sediments is usually controlled by the textural characteristics of the sediments, the composition of material supplied to the depositional environment and by the physical-chemical modifications of the material during and after deposition.

In the Gulf of St. Lawrence and Estuary the grain size of the sediments decreases with increasing water depths. Fine-grained sediments referred to as muds and sandy muds (containing > 70-95 % of material < 53 μm) occur at the mouth of the St. Lawrence Estuary and in the central part of the submarine troughs and shelf valleys whereas sandy sediments cover adjacent shelf areas and upper slopes of the troughs (Nota and Loring 1964). The fine-grained sediment–water interface in the troughs is oxidizing and is characterized by a thin (0.2-2 cm) brown layer enriched in Fe^{+3} and Mn. Below the interface, reducing conditions prevail and are characterized by dark greenish grey sediments in which Fe^{+2} and negative E_h conditions predominate (Loring and Nota 1968). Dynamic interactions across this interface during particulate resuspension apparently result in significant exchanges of Mn and Cd across this boundary (Yeats et al. 1979; Sundby and Silverberg 1985; Gendron et al. 1986). Most of the sedimentary material has been derived from the glacial erosion of the crystalline rocks of the Canadian Shield. This has produced large quantities of physically comminuted but chemically undecomposed material of a wide range of grain sizes. This rock flour, which contains some or all of the original igneous minerals with their inherited heavy metals, has been transported to and deposited in the Gulf (Loring and Nota 1973). Mineralogically, the sand and silt size material is characterized by high concentrations of plagioclase feldspars (average: 44 %) with lesser amounts of quartz (30 %), potash feldspars (16 %), and heavy minerals (10 %). The fine-grained clay size (< 2 μm) material contains chlorite, illite, kaolinite as well as plagioclase feldspar, pyroxenes, amphiboles, and quartz.

The trace metal composition of the Gulf sediments reflects their provenance. Table VII.1 shows the average and range of concentrations for 219 sediment samples from the St. Lawrence system. Compared with the earth's crust, igneous rocks, and shales (Table VII.1), the average trace metal content of the St. Lawrence sediments most closely resembles the trace metal composition of the earth's crust in general and of granodioritic rocks in particular. This is consistent with the bulk composition of the Canadian Shield whence most of the modern sediments were initially derived (Nota and Loring 1964), except for elevated average Hg concentrations resulting from industrial contamination (Loring 1975).

Statistical analysis indicates that the abundance and distribution of trace metals in the Estuary and Gulf is directly ($r = 0.74$ for Cr, $r = 0.75$ for Cu, $r = 0.22$ for Hg, $r = 0.84$ for Li, $r = 0.33$ for Pb, $r = 0.64$ for Ni, $r = 0.59$ for V, and $r = 0.67$ for Zn: $n = 219$) related to the sedimentation pattern of natural and anthropogenic fine grained (< 53 μm) inorganic and organic material.

Regionally (Fig. VII.2), high concentrations of total Zn (> 100 mg·kg^{-1}), Cu (> 30 mg·kg^{-1}), Pb (> 20 mg·kg^{-1}), Co (> 15 mg·kg^{-1}), Ni (> 30 mg·kg^{-1}), Cr (> 70 mg·kg^{-1}), and V (> 80 mg·kg^{-1}) occur in fine-grained sediments, at the

TABLE VII.1. Average abundance of trace elements in rock types and St. Lawrence sediments. (All values in mg·kg^{-1}).

	Earth's Crust	Ultra-mafic	Basalt	Grano-diorites	Granite	Shale	St. Lawrence Mean n = 219	St. Lawrence Range
As	1.8	1	2	2	1.5	15	6	1- 36
Be	2.8	—	0.5	2	5	3	2.2	1.0 - 3.0
Cd	0.2	—	0.2	0.2	0.2	0.2	0.21	0.04- 0.87
Co	25	150	50	10	1	20	12	3- 22
Cr	100	2000	200	20	4	100	70	8- 241
Cu	55	10	100	30	10	50	19	3- 76
Hg	0.08	—	0.08	0.08	0.08	0.5	0.46	0.10- 12.3
Li	20	—	10	25	30	60	26	10- 35
Ni	75	2000	150	20	0.5	70	27	4- 160
Pb	12.5	0.1	5	15	20	20	21	8- 66
V	135	50	250	100	20	130	92	4- 168
Zn	70	50	100	60	40	25	69	8- 215

mouth of the St. Lawrence Estuary, and in the central part of the submarine troughs and shelf valleys of the open Gulf. The lowest concentrations of these elements occur in the sandy shelf sediments of the open Gulf and Estuary. Local high anomalies of Ni (150-160 mg·kg^{-1}), Cr (100-241 mg·kg^{-1}), and V (159-168 mg·kg^{-1}), however, occur in the inshore sediments of the Bay of Islands, Newfoundland, most likely as a result of the seaward dispersal of detrital Ni, Cr, and V bearing minerals from nearby ultra-basic rocks (Loring 1979).

In the Saguenay Fjord, the highest concentrations of Zn, Cu, Pb, As, Cd, Co, Ni, Cr, and V occur in the fine-grained sediments in the upper part of the Fjord and lowest concentrations occur in the sandy sediments of the lower reaches (Fig.VII.2a-g). Mercury has a very distinctive distribution pattern (Fig.VII.3, see also Fig.VIII.2) with the highest concentrations (> 10 mg·kg^{-1}) occurring at the head of the Saguenay Fjord and decreasing seaward through the Fjord into the Lower Estuary and open Gulf.

Since the trace metals for the most part show comparable distributions, they can be combined into one distribution pattern (Fig. VII.4) that shows the relative distribution of the high, medium, and low concentrations in the region. The general pattern reflects their close relationship to sediment texture and morphology with high concentrations occurring in the fine-grained sediments of the submarine troughs and shelf valleys and the lowest in the sandy shelf sediments. The relative enrichment of an individual metal at specific locations is shown on the illustration by its chemical symbol. Figure VII.4 indicates that the sediments at the head of the Saguenay Fjord are relatively enriched in Hg, Zn, Pb, and As whilst those at the mouth of the St. Lawrence River are enriched in Zn, Pb, and Cu. Mercury is also enriched in the sediments of the Lower Estuary. Local anomalies of Hg as well as Cd occur in Chaleur Bay sediments, and Cr, Ni and V anomalies occur in the Bay of Islands (Newfoundland) sediments.

Since the total metal concentrations vary directly with grain size, it is obvious that valid regional comparisons can only be made between texturally equivalent sediments: e.g. the total Zn concentrations in the muds (sediments containing > 70 % by weight

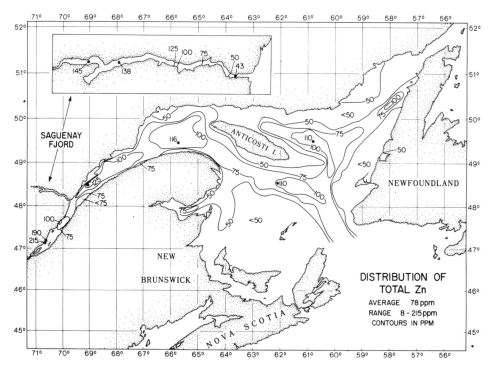

FIG. VII.2a. Distribution of total Zn in surface sediments (from Loring 1978).

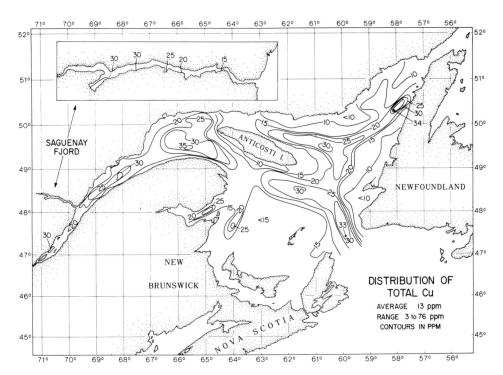

FIG. VII.2b. Distribution of total Cu in surface sediments (from Loring 1978).

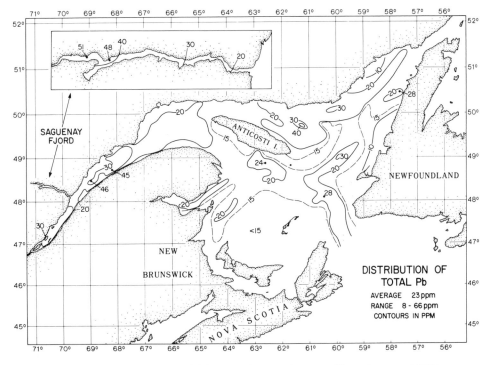

FIG. VII.2c. Distribution of total Pb in surface sediments (from Loring 1978).

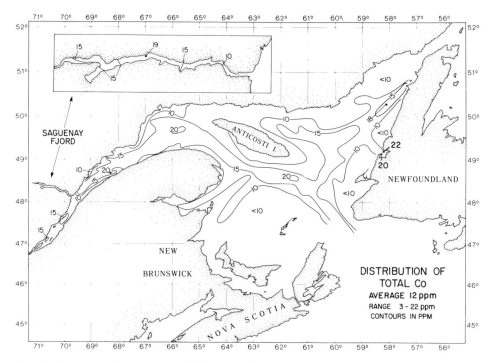

FIG. VII.2d. Distribution of total Co in surface sediments (from Loring 1979).

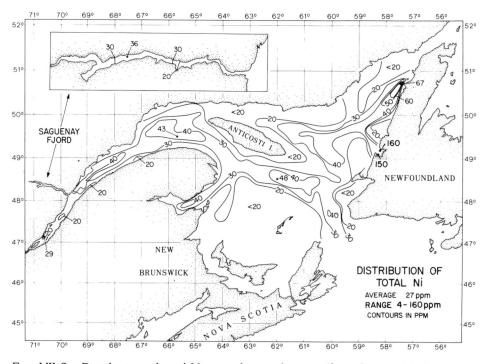

FIG. VII.2e. Distribution of total Ni in surface sediments (from Loring 1979).

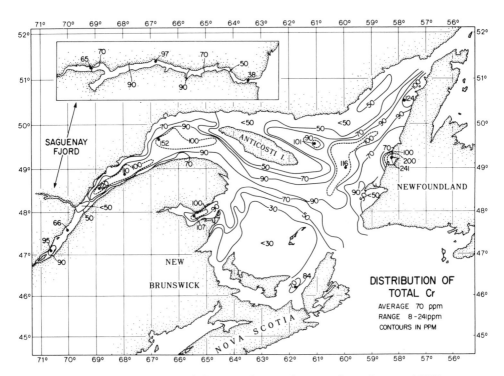

FIG. VII.2f. Distribution of total Cr in surface sediments (from Loring 1979).

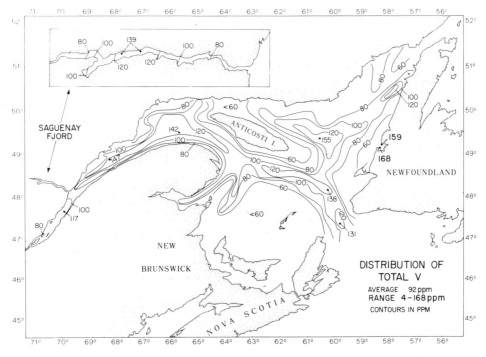

FIG. VII.2g. Distribution of total V in surface sediments (from Loring 1979).

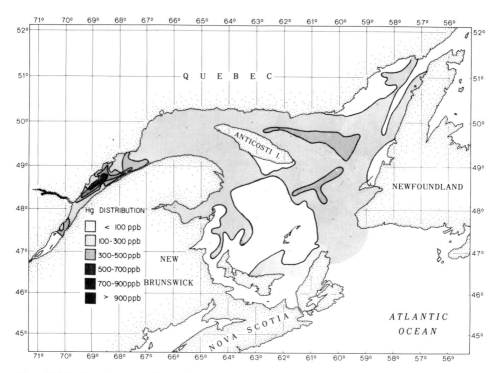

FIG. VII.3. Distribution of total Hg in surface sediments (from Loring 1975).

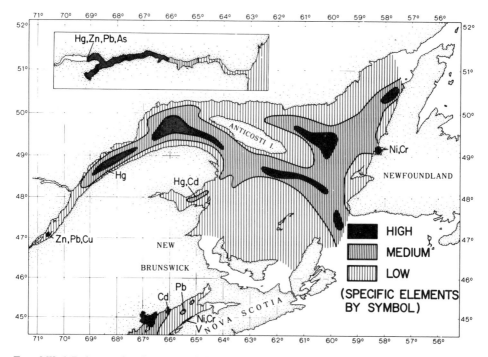

FIG. VII.4. Relative distributions of the high, medium, and low concentrations of Zn, Cu, Pb, Ni, Cr, V, As, and Cd in surface sediments (from Loring 1981, with permission).

material $< 53 \mu$m) of the Lower Estuary should only be directly compared with total Zn concentrations in the muds from the open Gulf.

Figure VII.5 shows the total concentrations of Zn, Cu, Pb, Co, Ni, Cr, Hg, As, and Cd in the fine-grained sediments of each region of the Gulf and those found in texturally equivalent sediments of the Bay of Fundy (Loring 1982), Placentia Bay, Newfoundland (Willey 1976), the Arctic region (Loring 1984) and some uncontaminated and contaminated rivers, estuaries and fjords of western Europe. From Fig. VII.5 it can be seen that average concentrations of Zn, Cu, and Pb are highest in the Upper Estuary and Saguenay Fjord and decrease seaward to the open Gulf. The average concentrations of Co, Ni, Cr, and Cd, however, vary very little between the different parts of the Gulf. There is, in contrast, a very strong gradient in mercury concentrations from the Saguenay Fjord sediments through the Lower Estuary to the open Gulf sediments. Arsenic is also relatively enriched in the Saguenay sediments but not elsewhere.

The metal concentrations of the open Gulf compare favourably to most metal concentrations found in the uncontaminated sediments from other areas of the Canadian east coast — the Bay of Fundy, Placentia Bay, and the Arctic region (Fig. VII.5). In the fine-grained sediments in the Gulf region, concentrations of Zn and Cu are higher than, and Pb, Co, and Ni concentrations are comparable to, levels found in the relatively uncontaminated sediments of Solway Firth. Concentrations of these elements in the Gulf of St. Lawrence are not, however, nearly as high as some of the concentrations reported for the contaminated sediments of the Firth of Clyde (MacKay et al. 1972: Pb, Ni), the Severn Estuary (Stoner 1974: all except Co), the Clyde Estuary (MacKay et al. 1972), the Wadden Sea (deGroot et al. 1974), and the heavily contaminated sediments of Sorfjord, Norway (Skei et al. 1972). Chromium levels in the Gulf are, however, higher than those of Solway Firth and the Firth of Clyde but much lower than those reported for the Clyde Estuary and the Rhine River. Arsenic levels in the Saguenay Fjord are comparable to those reported for the Wadden Sea, but much lower than those of the Rhine River. Cadmium levels are much less than those reported for the Severn Estuary, the Firth of Clyde, the Clyde Estuary, and the Wadden Sea. Mercury levels for all parts of the Gulf are also lower on the average than those reported for the Wadden Sea but those of the Saguenay Fjord reach comparable values to the Rhine River sediments.

Together the data indicate that, except for Hg, metals in the St. Lawrence sediments have not yet reached, and may not reach, the relatively high levels attained in the contaminated sediments of certain European estuaries. Although detectable, the industrial contributions of Zn, Cu, Pb, Cr, V, As, and Cd to the St. Lawrence sediments are comparatively low at the present time. Continued industrial inputs may lead, however, to further metal contamination in the future, particularly in the sediments of the Upper St. Lawrence Estuary and Saguenay Fjord. Unfortunately, industrial activity has already resulted in serious mercury contamination of the Saguenay Fjord (Loring 1975; Loring and Bewers 1978; Smith and Loring 1981). (See Chapter VIII).

Transport and Deposition Modes of the Trace Metals

Trace metals have been, and are being, introduced into the Estuary and Gulf in solution and in association with fine-grained solid and colloidal inorganic particles (see Chapter VI). Measurements of the total metal concentrations are, however, a poor means of determining the carriers and transport modes of these particulate metals because part of the metal load is loosely bound to the particles and part is locked up physically and/or chemically in detrital particles and minerals.

Trace Metal Partition

Selective chemical methods have been developed to partition the particulate and sedimentary metals into their loosely bound and residual phases (Hirst 1962; Loring and Not 968; Loring 1976a, b). Such a chemical partition, using a weak acetic acid

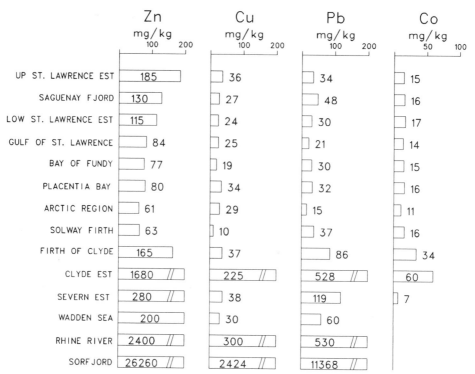

	Zn mg/kg	Cu mg/kg	Pb mg/kg	Co mg/kg
UP ST. LAWRENCE EST	185	36	34	15
SAGUENAY FJORD	130	27	48	16
LOW ST. LAWRENCE EST	115	24	30	17
GULF OF ST. LAWRENCE	84	25	21	14
BAY OF FUNDY	77	19	30	15
PLACENTIA BAY	80	34	32	16
ARCTIC REGION	61	29	15	11
SOLWAY FIRTH	63	10	37	16
FIRTH OF CLYDE	165	37	86	34
CLYDE EST	1680	225	528	60
SEVERN EST	280	38	119	7
WADDEN SEA	200	30	60	
RHINE RIVER	2400	300	530	
SORFJORD	26260	2424	11368	

FIG. VII.5a. Concentrations of total Zn, Cu, Pb, and Co in fine sediments from different sectors of the Gulf of St. Lawrence, and a number of other eastern Canadian and European locations.

	Ni mg/kg	Cr mg/kg	Hg mg/kg	As mg/kg	Cd mg/kg
UP ST. LAWRENCE EST	27	92	0.38	6	0.26
SAGUENAY FJORD	28	83	3.60	21	0.25
LOW ST. LAWRENCE EST	33	99	0.44	7	0.22
GULF OF ST. LAWRENCE	36	87	0.22	6	0.26
BAY OF FUNDY	22	73	0.06	9	0.24
PLACENTIA BAY	41	53			
ARCTIC REGION	22	64	0.05		0.16
SOLWAY FIRTH	38	35			2.90
FIRTH OF CLYDE	50	64		8	3.40
CLYDE EST	69	624		22	7
SEVERN EST	36	71	0.15	15	1.90
WADDEN SEA	21	100	1.00	30	1.00
RHINE RIVER	48	640	10	200	13
SORFJORD					121

FIG. VII.5b. Concentrations of total Ni, Cr, Hg, As, and Cd in fine sediments from different sectors of the Gulf of St. Lawrence, and a number of other eastern Canadian and European locations.

extraction, allows certain deductions to be made on the site of an element and the pathways by which it has entered the sediments.

The acetic acid soluble fraction is operationally referred to as the loosely bound or non-detrital fraction (Hirst 1962; Loring 1976a, b; 1978; 1979; 1981). It is believed to represent some proportion of the element that was initially leached from the source rocks and/or supplied in dissolved form from industrial sources. This part of the total metal content is incorporated into the particulate phase both at the site of weathering and during transport. Non-detrital metals are also transferred from solution or colloidal form to the sedimentary phase during deposition by precipitation, adsorption onto SPM and by extraction by organisms. Some of the non-detrital contributions may be derived from ions released during diagenetic alterations, from metals held in carbonates, easily soluble amorphous compounds of Fe and Mn, and/or held in surface and interstitial positions of the particles.

In contrast, that part of the total metal concentration that is held in the acid insoluble phase, operationally defined as the residual or detrital phase, is believed to represent the fraction of the metal content held in the lattices of detrital silicate, oxide, and sulphide minerals and secondary insoluble compounds, mainly comprising oxides of iron and manganese. These minerals and compounds are transported as fine clastic particles and deposited along with other detrital material of comparable grain size and settling rates.

In the St. Lawrence sediments, most of the total trace metal concentrations (61-98%) resides in the detrital (acid insoluble) phase, and a smaller (2-39%), but geochemically significant, fraction of the total resides in the non-detrital (acid soluble) fraction (Table VII.2). A notable exception, however, is cadmium which occurs predominantly (39-80%) in the non-detrital fraction.

The partition data provide valuable information on the geochemical behaviour of each element and the location and extent of contamination in the region. The percentages of the total metal concentrations held in the non-detrital fractions ($Me_{nd}/Me_t \times 100$), and the non-detrital and detrital concentrations vary both regionally as well as with changes in sediment texture in different parts of the region. These variations are illustrated in Fig. VII.6-8. The non-detrital and detrital concentrations ($mg \cdot kg^{-1}$) of Zn (Fig. VII.6), for example, and most of the other elements, usually increase significantly ($P < 0.01$) with decreasing grain size of the sediments. The main exceptions to this were found to be for non-detrital Pb (Lower Estuary and open Gulf), and detrital Pb (Upper Estuary — Fig. VII.7). Detrital V (Fig. VII.8b) is also an exception in that it decreases with decreasing grain size in the upper estuarine sediments.

TABLE VII.2. Non-detrital metal concentrations as a percentage of total metal concentration in sediments from the Gulf of St. Lawrence.

Metal	St. Lawrence River	Upper Estuary	Saguenay Fjord	Lower Estuary	Open Gulf	Overall
Zn	41[c]	17 –39[c]	14[a] –29[c]	18[b] –20[a,b,c,d]	8[c] – 9[a,b,d]	8 – 41
Cu	30[c]	17[a,c]–20[b]	14[b] –21[c]	13[c] –20[a]	7[b] –13[a]	7 – 30
Pb	<24[c]	17[a] –26[c]	12[d] –25[c]	15[b] –19[d]	18[c] –25[c]	12 – 26
Co	23[c]	14[a,b]–20[c]	8[a] –25[b]	18[b,c]–24[b]	8[c] –13[a,d]	8 – 25
Ni	—	13[a,b]–19[c]	11[a] –29[b]	16[c,d]–23[a]	12[a,b] –15[c,d]	11 – 29
Cr	—	4[a] –11[c]	2[a] – 9[d]	2 – 5[d]	2[a,c,d]– 3[b]	2 – 11
V	—	5[a] –10[c]	6[a] –23[d]	6[a] –18[d]	8[c,d] –11[b]	5 – 23
Cd	—	67[c] –80[a]	39[a,c]–71[b,d]	74[c] –76[d]	—	—
No. of samples	2	20	19	49	127	217

Sediment type:
[a] Muddy sand: 5-30% pelite (< 0.05 mm), > 70% sand (2—0.05 mm).
[b] Very sandy pelites (muds): > 30% sand, > 30% pelite.
[c] Sandy pelites (muds): 5-30% sand, > 70% pelite.
[d] Pelites (muds): < 5% sand, > 95% pelites.

109

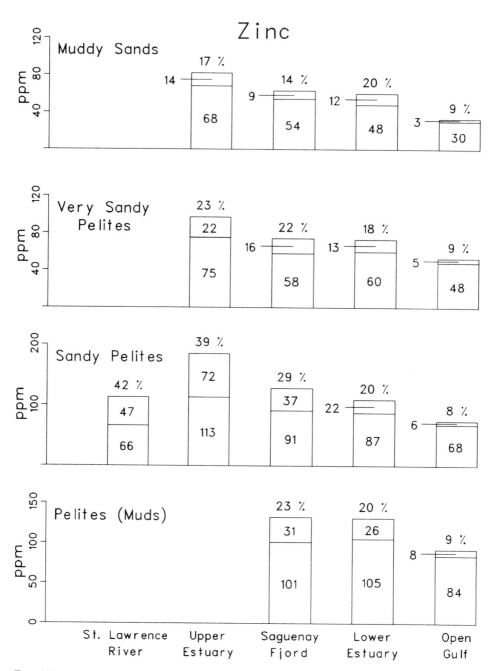

FIG. VII.6. Partition of total Zn into its non-detrital (number in top part of bar) and detrital components (bottom number) in the various sediment types in the Gulf of St. Lawrence. The non-detrital/total Zn percentage is indicated on the top of each bar (redrawn from Loring 1981, with permission).

Regionally the highest concentrations of detrital and non-detrital Zn, Cu, and Pb occur in the fine-grained sediments of the Upper Estuary and Saguenay Fjord and decrease through the Lower Estuary to the open Gulf. Non-detrital and detrital Co concentrations are, however, relatively constant between the different parts of the system. Non-detrital Cr is highest in the Upper Estuary muds and decreases seaward to

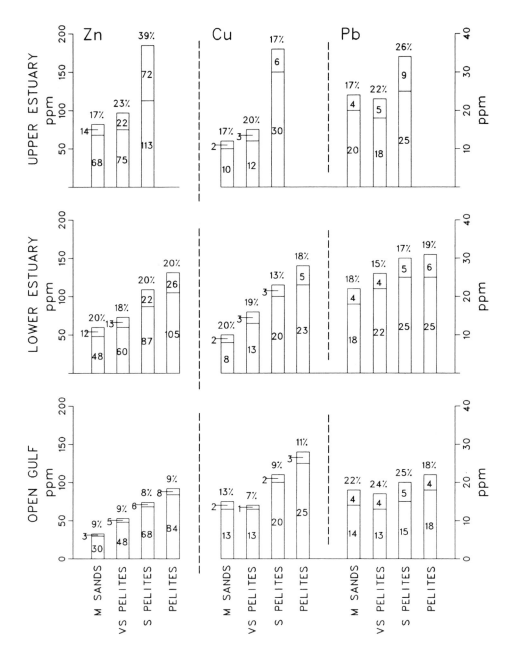

FIG. VII.7. Partition of total Zn, Cu, and Pb into their average non-detrital (number in upper part of bar) and detrital (number in lower bar) components in different sectors and sediment types in the Gulf of St. Lawrence. Number on top of the bar indicates the percentage of the total metal found in the non-detrital component (redrawn from Loring 1978, with permission).

the open Gulf. In contrast, non-detrital V is highest in the Saguenay Fjord and Lower Estuary muds.

Significant textural and regional changes also occur in Me_{nd}/Me_t ratios (Fig. VII.6-8). The ratio increases with decreasing grain size for Zn, Pb, Ni, Cr, and V in the Upper Estuary, but remains relatively constant with grain size for Co and Cu in the Upper Estuary, for Zn, Cu, and Pb in the Lower Estuary, and for Zn, Cu, Co, Ni, Cr, and

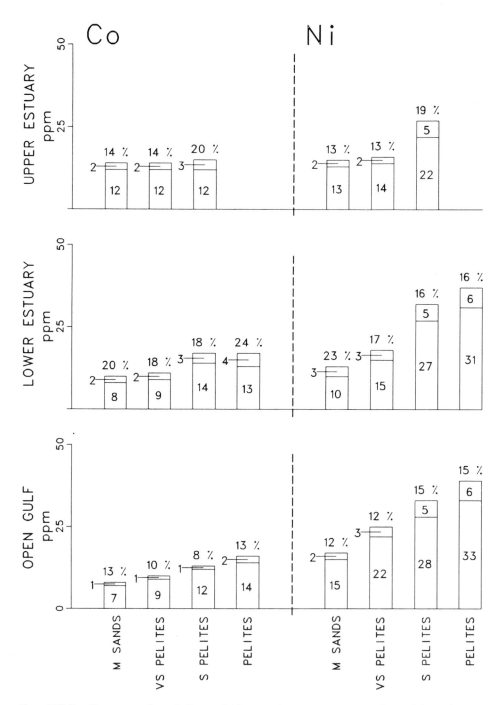

FIG. VII.8a. Partition of total Co and Ni into their average non-detrital (number in upper part of bar) and detrital (number in lower bar) components in different sectors and sediment types in the Gulf of St. Lawrence. Number on top of the bar indicates the percentage of the total metal found in the non-detrital component (redrawn from Loring 1979, with permission).

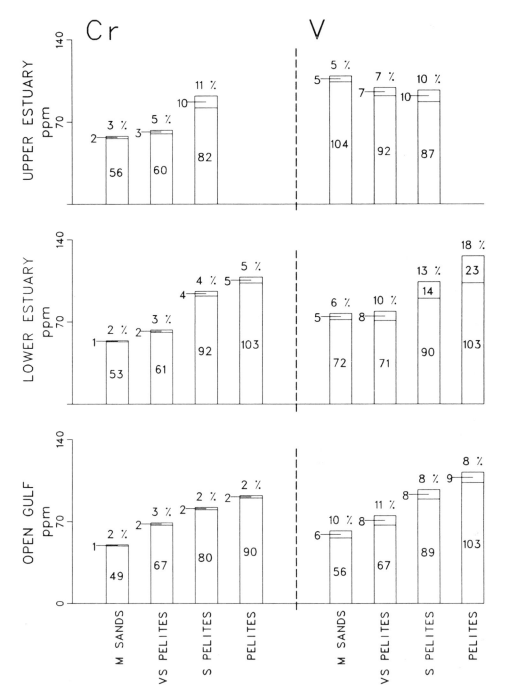

FIG. VII.8b. Partition of total Cr and V into their average non-detrital (number in upper part of bar) and detrital (number in lower bar) components in different sectors and sediment types in the Gulf of St. Lawrence. Number on top of the bar indicates the percentage of the total metal found in the non-detrital component (redrawn from Loring 1979, with permission).

V in the open Gulf. In the Saguenay Fjord, the highest ratios occur for Zn, Cu, and Pb in the sandy muds at the head of the Fjord and the lowest ratios for elements occur in the muddy sands. Regionally, there is a significant decrease in the Me_{nd}/Me_t ratio for texturally equivalent sediments (sandy muds and muds) from the St. Lawrence River (Zn and Cu) and Upper Estuary (Zn, Cu, Pb, and Cr) and from the Saguenay Fjord seaward through the Lower Estuary to the open Gulf.

As a result, the relatively high total concentrations of Zn, Cu, and Pb in the Upper Estuary and Saguenay Fjord can be directly attributed to the high proportion of the non-detrital component and the seaward decrease in total elemental concentrations to a decrease in the non-detrital concentrations in the sediments. It indicates (Loring 1978; 1979; 1981) that there has been a relatively high removal of the non-detrital component to the sediments at the mouth of the St. Lawrence and Saguenay rivers where relatively high levels of metals are introduced, in dissolved and solid phases, into the estuarine environment from natural and anthropogenic sources. Most of the non-detrital component is apparently removed in the estuarine environment, but a small amount escapes seaward to be deposited in decreasing quantities as the distance from the sources increases. This is illustrated by non-detrital Zn which declines seaward from 39% of the total in the Upper Estuary to 8% of the total in the open Gulf (Fig. VII.6). The constant Me_{nd}/Me_t ratio for the elements (except for V) with textural changes in the Lower Estuary and open Gulf implies that only small additional amounts of the metals are removed from solution by SPM in a variety of ways during transit through the Lower Estuary to the open Gulf or by adsorption onto sediment surfaces following deposition.

Host Mineral / Compound Carriers of Detrital Trace Metals

Fine-grained host minerals/compounds are the main carriers of the detrital heavy metals (69-89% of the total) and control the concentrations of most metals, except for that of Cd, in the sediments. The highest detrital metal concentrations occur along with other fine-grained material in the submarine troughs and shelf valleys where they have been deposited in response to the present sedimentological environment.

Statistically (Tables VII.3-6), a significant ($P < 0.01$) positive correlation of most detrital metals with the amounts of fine grained mud (material < 63 μm) shows the strong increase in concentration of detrital metals with decreasing grain size. The significant ($P < 0.01$) inverse correlation of the detrital metals with Si shows that quartz (SiO_2) acts as a diluent of the metal-bearing minerals in the sediments. In many sedimentary environments, most trace metals covary with Al and grain size because fine-grained aluminosilicates are the main host minerals for the metals. This is not the case in the St. Lawrence region where no significant ($P \leq 0.05$) relationships occur between Al and Zn, Cu, Pb, Co, and V (Table VII.3 and VII.6). In the open Gulf there is a significant positive covariance of the detrital metals with Al that would reflect a general increase in metal concentration with increasing concentrations of aluminosilicate minerals as the grain size of the sediments decreases.

The absence of the expected relationship is due to the relatively high amounts of Al bearing feldspars in the sand (40-60% feldspars) as well as in the mud (25-40% feldspars) size material, particularly in the northern Gulf regions. These feldspars contribute Al but little or no trace metals to the sediments and so mask the detrital metal/Al relationship. In contrast, strong covariances of the detrital metals with Li, which does not enter the feldspar lattices, as well as with Mg and Fe (Tables VII.3-6) in all regions, argue that these metals reside primarily in the ferromagnesian minerals: biotite mica, amphiboles, chlorite, pyroxenes, and garnets and other discrete oxide and sulphide minerals associated with the heavy mineral fraction of the sediments.

Figure VII.9 illustrates the close covariance of Zn with Li but not with Al concentrations as the metals increase with decreasing size fractions in 18 samples from different parts of the Gulf. The increase in metal concentrations with decreasing grain size reflects the granulometric changes in the particle size of their host minerals whereas

TABLE VII.3. Correlation matrices for detrital (D) and non-detrital (ND) Zn, Cu, and Pb in sediments of the Upper St. Lawrence Estuary and the Saguenay Fjord. This table is extracted from more complete analyses (Loring 1976a, 1978).

	Zn		Cu		Pb	
	D	ND	D	ND	D	ND
UPPER ST. LAWRENCE ESTUARY						
Mud (%)	0.84	0.87	0.88	0.74	—	0.80
Organic Carbon	0.90	0.96	0.98	0.77	—	0.93
Carbonate	0.87	0.90	0.88	0.77	—	0.84
Si	−0.73	−0.82	−0.84	−0.74	—	−0.80
Al	—	—	—	—	—	0.58
Fe (total)	—	—	—	—	—	—
Fe (non-det.)	0.80	0.85	0.82	0.82	—	0.82
Mg	0.67	0.75	0.80	0.63	0.54	0.72
Li (total)	0.72	0.82	0.84	0.73	—	0.81
		(n = 20, P < 0.01 for r > 0.56)				
SAGUENAY FJORD						
Mud (%)	0.95	0.88	0.97	0.74	0.79	—
Organic Carbon	0.81	0.95	0.89	0.85	0.73	0.62
Carbonate	0.69	—	—	—	0.76	—
Si	−0.86	−0.83	−0.91	−0.84	−0.62	—
Al	—	—	—	—	—	—
Fe (total)	0.74	0.57	0.72	—	0.78	—
Fe (non-det.)	0.59	0.55	0.69	—	—	—
Mg	0.92	0.82	0.94	0.67	0.77	—
Li (total)	0.94	0.70	0.90	0.58	0.72	—
		(n = 20, P< 0.01 for r > 0.56)				

TABLE VII.4. Correlation matrices for detrital (D) and non-detrital (ND) Zn, Cu, and Pb in sediments of the Lower St. Lawrence Estuary and the open Gulf of St. Lawrence. This table is extracted from more complete analyses in Loring (1978).

	Zn		Cu		Pb	
	D	ND	D	ND	D	ND
LOWER ST. LAWRENCE ESTUARY						
Mud (%)	0.78	0.63	0.63	0.40	0.42	—
Organic Carbon	0.81	0.74	0.65	0.38	0.43	0.47
Carbonate	0.62	0.68	0.54	0.62	—	0.36
Si	−0.75	−0.65	−0.60	−0.49	—	—
Al	—	—	—	—	—	—
Fe (total)	0.66	0.41	0.48	—	—	—
Fe (non-det.)	0.78	0.87	0.64	0.59	—	0.47
Mg	0.67	0.37	0.50	—	—	—
Li (total)	0.84	0.73	0.68	0.55	0.40	0.40
		(n = 49, P < 0.01 for r > 0.36)				
OPEN GULF OF ST. LAWRENCE						
Mud (%)	0.82	0.38	0.48	—	0.29	—
Organic Carbon	0.78	0.25	0.53	—	0.30	—
Carbonate	—	—	—	—	−0.22	0.32
Si	−0.42	−0.26	−0.24	—	—	−0.45
Al	0.72	0.42	0.41	—	0.46	—
Fe (total)	0.87	0.46	0.50	—	0.42	—
Fe (non-det.)	0.25	0.58	—	—	0.27	—
Mg	0.68	0.26	0.50	—	—	—
Li (total)	0.92	0.35	0.61	—	0.29	—
		(n = 130, P< 0.01 for r > 0.22)				

TABLE VII.5. Correlation matrices for detrital (D) and non-detrital (ND) Co, Ni, Cr and V in sediments of the Upper St. Lawrence Estuary and the Saguenay Fjord. This table is extracted from more complete analyses (Loring 1976b, 1979).

	Co		Ni		Cr		V	
	D	ND	D	ND	D	ND	D	ND
UPPER ST. LAWRENCE ESTUARY								
Mud (%)	—	—	0.72	0.69	0.82	0.86	—	0.64
Organic Carbon	—	0.58	0.73	0.79	0.84	0.94	—	—
Carbonate	—	—	0.62	0.71	0.81	0.89	—	—
Si	—	−0.61	−0.72	−0.65	−0.79	−0.84	—	−0.68
Al	—	—	—	—	0.66	0.55	−0.63	0.77
Fe (total)	—	—	—	—	—	—	0.55	—
Fe (non-det.)	—	0.64	0.69	0.71	0.74	0.83	—	0.65
Mg	—	—	0.80	—	0.91	0.80	—	0.69
Li (total)	—	—	0.70	0.67	0.94	0.86	—	0.74
(n = 20, P < 0.01 for r > 0.56)								
SAGUENAY FJORD								
Mud (%)	0.61	—	0.90	—	0.55	0.90	—	0.76
Organic Carbon	—	—	0.70	—	—	0.71	—	—
Carbonate	—	—	0.61	—	—	0.55	0.67	—
Si	−0.60	—	−0.88	—	−0.64	−0.91	—	−0.81
Al	—	—	—	—	—	—	—	—
Fe (total)	0.83	—	0.74	—	0.69	0.65	0.62	0.66
Fe (non-det.)	—	—	0.64	—	—	0.65	—	0.71
Mg	0.66	0.57	0.88	0.54	0.57	0.87	—	0.80
Li (total)	0.70	—	0.95	—	0.68	0.93	—	0.88
(n = 20, P < 0.01 for r > 0.56)								

TABLE VII.6. Correlation matrices for detrital (D) and non-detrital (ND) Co, Ni, Cr and V in sediments of the Lower St. Lawrence Estuary and the open Gulf of St. Lawrence. This table is extracted from more complete analyses in Loring (1979).

	Co		Ni		Cr		V	
	D	ND	D	ND	D	ND	D	ND
LOWER ST. LAWRENCE ESTUARY								
Mud (%)	0.56	0.55	0.81	0.68	0.78	0.72	0.50	0.60
Organic Carbon	0.63	0.52	0.83	0.76	0.84	0.80	0.58	0.62
Carbonate	—	0.66	0.69	0.69	0.58	0.66	—	0.41
Si	−0.58	−0.57	−0.85	−0.77	−0.76	−0.72	−0.47	−0.65
Al	—	—	—	—	—	—	—	—
Fe (total)	0.58	—	0.68	0.59	0.66	0.38	0.50	0.38
Fe (non-det.)	0.46	0.70	0.80	0.78	0.74	0.87	0.52	0.59
Mg	0.71	—	0.75	0.56	0.76	0.42	0.50	0.40
Li (total)	0.70	0.61	0.92	0.82	0.89	0.77	0.66	0.59
(n = 49, P < 0.01 for r > 0.36)								
OPEN GULF OF ST. LAWRENCE								
Mud (%)	0.72	0.28	0.54	0.62	0.70	0.29	0.72	—
Organic Carbon	0.65	—	0.59	0.67	0.66	0.28	0.70	—
Carbonate	—	—	—	—	—	—	—	—
Si	−0.34	—	—	−0.34	—	−0.68	−0.31	−0.31
Al	0.68	0.24	0.32	0.41	0.58	—	0.72	—
Fe (total)	0.80	0.26	0.51	0.59	0.72	—	0.81	—
Fe (non-det.)	0.37	0.33	—	0.26	0.25	0.30	0.24	—
Mg	0.66	—	0.74	0.62	0.78	—	0.62	—
Li (total)	0.82	0.36	0.67	0.80	0.75	—	0.83	—
(n = 130, P < 0.01 for r > 0.22)								

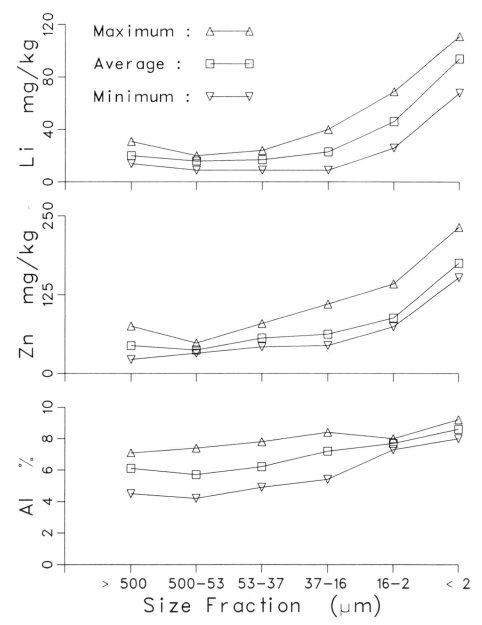

FIG. VII.9. Change of Al, Zn, and Li concentrations with grain size in 18 samples from different parts of the Gulf of St. Lawrence.

the spread of the metal concentrations in each size fraction reflects mineralogical differences due to the provenance of the material. For example, the relatively small changes in Al concentrations with large changes in the size fractions reflects the presence of relatively large amounts of Al-bearing feldspars in all size fractions, whereas the spread of Al concentrations in each size fraction, particularly the fractions > 16 μm, reflect the mineralogical differences between the northern and southern Gulf (Loring and Nota 1973). The close covariance of Li and Zn with changes in the size fractions reflects the increase in the Li and Zn bearing minerals with decreasing grain size and the spread within each fraction reflects the effects of provenance on the Li and Zn distribution.

Microprobe analyses (Loring 1976a, b; 1978) show that discrete grains of sulphide minerals (galena, sphalerite, pyrite, and chalcopyrite) are present in sufficient quantities to account for most of the detrital Zn, Cu, and Pb in the sediments and that discrete grains of chromite, magnetite, and illmenite along with the sulphide minerals and the ferromagnesian mineral hornblende, which is rich in Ni, account for most of the Cr, V, Co, and Ni in the sediments.

A correlation of detrital Zn and Pb with carbonate (expressed as CO_2) in some regions appears to be an indirect relationship in which carbonate concentrations parallel the increase in detrital element concentrations with decreasing grain size. A positive correlation of detrital Zn with carbonate in the Saguenay sediments may, however, be more direct and reflect the mineralogical association of zinc spinels with metamorphic limestone (skarn) particles of detrital origin. Although organic carbon appears to vary directly with detrital Zn, Cu, and Pb, the relationship appears to be secondary rather than primary. This is demonstrated by extraction of the acetic acid insoluble residues with H_2O_2, which removes organically bound metals. The data show that no significant amounts of the metals are dissolved from the detrital fraction. Instead, the relationship derives from a grain size relationship in which the amounts of organic matter and the detrital metal-bearing minerals increase with decreasing grain size.

Carriers of Non-detrital Trace Metals

Fine-grained organic and inorganic particulate matter provide large surface areas for the adsorption of elements from solution and suitable media for element retention during transport and deposition. In most parts of the region, the comparatively high non-detrital element concentrations in the fine-grained sediments (Fig. VII.6-8) and their high covariances with mud (material $< 53 \mu$m) and organic matter (Tables VII.3-6), demonstrate that fine-grained material is closely associated with the abundance and distribution of the non-detrital metals.

The small quantities of metals in this fraction, however, make it difficult to determine their host minerals/compounds other than by a statistical analysis. In the inorganic fraction of most sediments, the high correlations of most metals with non-detrital iron (Table VII.3-6) suggest that such contributions are derived from iron oxide grain coatings, a phase that has often been observed elsewhere (Jenne 1968). Lesser correlations of the non-detrital contributions with Mg, Li, and carbonate also suggest that some contributions derive from ion-exchange positions on ferromagnesian minerals and from carbonate minerals. The organic fraction, depending on its nature, appears to be an important carrier of non-detrital metals in certain parts of the region. In Upper Estuary and Saguenay sediments, organic matter, mainly of terrestrial origin, appears to be the main carrier of non-detrital Zn, Cu, Pb, and Cr as well as Hg, and to a lesser extent Co, Ni, and V. In contrast, organic matter of marine origin in the open Gulf sediments does not appear to be a significant carrier of the non-detrital metals.

Dispersal Factors

A number of physical, chemical, hydrological and sedimentological factors control the abundance and distribution of the heavy metals in the sediments of the Gulf of St. Lawrence region (Loring 1975, 1976a, b, 1978, 1979, 1981). Statistical analyses of the interrelationships between the non-detrital and detrital concentrations of trace metals and other characteristics of the sediments have allowed certain deductions to be made about the factors that control the abundance and distribution of the elements. These characteristics include organic carbon and carbonate contents, the major elements, and non-detrital iron and manganese. Table VII.7 summarizes the role played by various sedimentary inorganic components and the carriers of the metals in the various parts of the St. Lawrence system.

Regionally, inputs of metals from natural and anthropogenic sources and their sedimentation rates, along with physical and chemical exchanges of particulate and

TABLE VII.7. Roles played by texture, inorganic and organic components, and/or the site of elements in the sediments.

Factor	Size	Role	Fraction	Zn	Cu	Pb	Co	Ni	Cr	V
Sand	> 50 μm	Diluent	D[a]	1,2,3[b]	1,2,3	1,2,3	2,3	1,2,3	1,2,3	2,3
			ND	1,2	1,2	1,2	3	1,2,3	1,2,3	1,2
		Concentrator	D							1
Mud	< 50 μm	Concentrator	D	1,2,3	1,2,3	1,2,3	2,3	1,2,3	1,2,3	2,3
			ND	1	1,2	1	2,3	1,2,3	1,3	2
Quartz	> 50 μm	Diluent	D	1,2,3	1,2,3		2,3	1,2	1,2	2
			ND	1,2	1,2	1,2	1,2	1,2,3	1,2,3	1,2,3
Aluminosilicates	< 50 μm	Companion	D	3	3	3	3	3	1,3	3
			ND	3		1	3	3	1	
		Diluent	D							1
Ferromagnesians	< 50 μm	Minor carrier	D	1,2,3	1,2,3		2,3	2,3	2,3	2,3
		Carrier	ND	1,2,3	1	1		2,3	2,3	2,3
Mg silicates	<50 μm	Carrier	D	2,3	2,3	1		1	1	1
			ND							
Chromite	< 50 μm	Major carrier	D						1,2,3	1,2,3
Magnetite	< 50 μm	Major carrier	D						1,2,3	1,3
Fe–Ti oxides	> 50 μm	Major carrier	D				3		3	1,3
		Minor carrier	ND						2	
Sulphides	< 50 μm	Major carrier	D	1	1	1	2,3	2,3		
		Minor carrier	D				1	1		
Fe oxide coating	< 50 μm	Minor carrier	D	1,2,3	1,2	3	3	1	1,3	3
		Major carrier	ND	1,2,3	1,2	1,2	1,2,3	1,2,3	1,2,3	1,2
Mn oxide coating	< 50 μm	Minor carrier	D	3	3	3	3	3	3	2,3
			ND							
Carbonates	< 50 μm	Companion	D	2,3	1,2	1,2	2,3	2,3	1,2	
			ND	1,2	1,2	1,2	2	1,2	1,2	2
Organic matter	< 50 μm	Companion	D	1,2,3	1,2,3	2,3	2,3	1,2,3	1,2,3	2,3
		Major carrier	ND	1,2	1,2	1,2	1	1,2	1	2
		Minor carrier	ND	3			2	3	2,3	

[a] D = detrital; ND = non-detrital
[b] 1 = Upper Estuary; 2 = Lower Estuary; 3 = open Gulf of St. Lawrence

119

dissolved metals in the water column (Chapter VI) and at the depositional site, control the abundance and distribution of trace metals.

Physically, fine particle size is the dominant controlling factor in most regions. Factor analyses (r-mode) indicate that the fine-grained sedimentation factor accounts for 40–46 % of the total problem variance in the St. Lawrence sediments (Loring 1978, 1979). This is because the detrital carriers are fine-grained and the small particles provide the most surface area for adsorption of metals from solution and for their retention in the non-detrital fraction during transport and deposition. This large surface area may also interact with interstitial waters following deposition.

The sedimentation of fine-grained clastic grains of metal-bearing silicate, oxide, and sulphide minerals and compounds has made the largest contribution (61–93 %) to total metal concentrations in the sediments. The concentrations are highest in the Estuary and they decrease seaward from the interior sources of terrigenous clastic material supplied to the Estuary by the St. Lawrence and Saguenay rivers. Local anomalies of detrital Cr, Ni, and V, however, are exceptions to this pattern: an enrichment of V in the sand-size material of the Upper Estuary results from the accumulation of vanadium-bearing minerals derived from the massive deposits of illmenite and magnetite at St. Urbain, Quebec; the local anomalies of Cr, Ni, and V off the west coast of Newfoundland result from the seaward dispersal of their host minerals from an ultrabasic rock complex near the Bay of Islands.

The concentrations of metals in the non-detrital fraction is related to grain size and chemical interactions within the system. The highest non–detrital concentrations occur in the fine-grained sediments deposited closest to the outflow of natural and anthropogenic dissolved and particulate matter and thereafter non-detrital concentrations decrease seaward.

The fine-grained sediments beneath the turbidity zone in the Upper Estuary appear to be the major sink for non-detrital Zn, Cu, Pb, Ni, Cr, and V (Loring 1978, 1979). This is a result of flocculation, deposition, and differential settling of relatively large particles enriched in organic matter (Kranck 1979) and available metals in the turbidity maximum. The organic matter, mostly of terrestrial origin (See Chapter V), acts as the main scavenger and means of transfer of available Cr, Zn, and Pb and to a lesser extent, Co, Ni, and Cu, to the sea floor. Hydrous iron oxides appear to be the most effective scavengers of Ni, V, and perhaps Cu and Co. A detailed study of the early diagenesis of Co at the sediment–water interface by Gendron et al. (1986) confirmed that the small amount of mobile Co is associated with Fe-Mn hydroxides. Some of these metal-rich materials may be resuspended and mixed upwards (Sundby and Silverberg 1985), but settling appears to be so rapid that there is little net loss of metals from the bottom. The high non-detrital Zn, Pb, and Cr to total metal ratios of the fine-grained sediments suggest that additional amounts of these metals are derived from solution to augment that already present in the particulate matter entering the Estuary. For most of the other metals such augmentation from solution is of lesser importance. These observations are consistent with measurements of trace metal distributions in the water column (see Chapter VI) which suggest that Zn and Cr are extensively removed from the water column in this region and that Co and Cu are removed to a lesser extent. (Water column measurements are not available for Pb).

Since the low number of particles escaping the Upper Estuary are mostly inorganic (Kranck 1979), there is a seaward decrease in the percentages of non-detrital metals held in the fine-grained sediments of the Lower Estuary. Particles escaping from the Saguenay Fjord and subsequently deposited with those of the Lower Estuary result in the enrichment of the sediment in anthropogenic non-detrital V, Zn, Cu, Pb, and Hg. In the Lower Estuary, most of the non-detrital Zn, Pb, Cu, Cr, Co, and Ni are probably associated with the particles on entry. Non-detrital V, however, appears to have travelled to the Lower Estuary in solution and then been adsorbed or become associated with hydrous iron oxides and to a lesser extent with hydrous Mn oxides. Only small quantities of the non-detrital metals are transported into the open Gulf through association with Fe/Mn oxides and in ion-exchange positions of the silicate minerals.

Although the actual contribution of the metals to the upper estuarine sediments is relatively small at the present time, when compared to some other contaminated estuaries of the world, the absolute quantities being supplied from natural and industrial sources may lead to excessive accumulation in the Upper Estuary and a continual input of the elements to seaward sediments in the future.

References

CAMERON, E. M. 1968. A geochemical profile of Swan Hills reef. Can. J. Earth Sci. 5: 287-309.

————— 1969. Regional geochemical study of the Slave Point carbonates, western Canada. Can. J. Earth Sci. 6: 247-268.

DeGROOT, A., W. SALOMONS, AND M.A. ALLERSMA. 1974. Process affecting heavy metals in estuarine sediments. Unpublished Report, Institute of Soil Fertility Haren (Gn), Netherlands.

FLANAGAN, F. J. 1973. 1972 values for international geochemical reference samples. Geochim. Cosmochim. Acta 37: 1189-1200.

GENDRON, A., N. SILVERBERG, B. SUNDBY, AND J. LEBEL. 1986. Early diagenesis of cadmium and cobalt in sediments of the Laurentian Trough. Geochim. Cosmochim. Acta 50: 741-747.

HATCH, W. R. AND W. L. OTT. 1968. Determination of sub-microgram quantities of mercury by atomic absorption spectrophotometry. Anal. Chem. 40: 2085-2087.

HIRST, D.M. 1962. The geochemistry of modern sediments from the Gulf of Paria — II The location and distribution of trace elements. Geochim. Cosmochim. Acta 26: 1147-1187.

JENNE, E. A. 1968. Controls on Mn, Fe, Co, Ni, Cu, and Zn concentrations in soils and water: the significant role of hydrous Mn and Fe oxides. Am. Chem. Soc. Adv. Chem. Ser. 73: 337-387.

KRANCK, K. 1979. Dynamics and distribution of suspended particulate matter in the St. Lawrence estuary. Nat. Can. (Que.) 106: 163-173.

LORING, D. H. 1975. Mercury in the sediments of the Gulf of St. Lawrence. Can. J. Earth Sci. 12: 1219-1237.

————— 1976a. The distribution and partition of zinc, copper, and lead in the sediments of the Saguenay Fjord. Can. J. Earth Sci. 13: 960-971.

————— 1976b. Distribution and partition of cobalt, nickel, chromium, and vanadium in the sediments of the Saguenay Fjord. Can. J. Earth Sci. 13: 1706-1718.

————— 1978. Geochemistry of zinc, copper, and lead in the sediments of the estuary and Gulf of St. Lawrence. Can. J. Earth Sci. 15: 757-772.

————— 1979. Geochemistry of cobalt, nickel, chromium, and vanadium in the sediments of the estuary and open Gulf of St. Lawrence. Can. J. Earth Sci. 16: 1196-1209.

————— 1981. Potential bioavailability of metals in eastern Canadian estuarine and coastal sediments. Rapp. P.V. Reun. Cons. Int. Explor. Mer 181: 93-101.

————— 1982. Geochemical factors controlling the accumulation and dispersal of heavy metals in the Bay of Fundy sediments. Can. J. Earth Sci. 19: 930-944.

————— 1984. Trace-metal geochemistry of sediments from Baffin Bay. Can. J. Earth Sci. 21: 1368-1378.

LORING, D. H., AND J.M. BEWERS. 1978. Geochemical mass balances for mercury in a Canadian Fjord. Chem. Geol. 22: 309-330.

LORING, D. H. AND D. J. G. Nota. 1968. Occurrence and significance of iron, manganese, and titanium in glacial marine sediments from the estuary of the St. Lawrence river. J. Fish. Res. Board Can. 25: 2327-2347.

————— 1973. Morphology and sediments of the Gulf of St. Lawrence. Bull. Fish. Res. Board Can. 182: 147 p.

LORING, D. H., AND R. T. T. RANTALA. 1977. Geochemical analyses of marine sediments and suspended particulate matter. Fish. Mar. Serv. Tech. Rep. 700: 58 p.

LORING, D. H., R. T. T. RANTALA, AND J.N. SMITH. 1983. Response time of Saguenay Fjord sediments to metal contamination. Environ. Biogeochem. Ecol. Bull. 35: 59-72.

MACKAY, D. W., W. HALCROW, AND I. THORNTON. 1972. Sludge dumping in the Firth of Clyde. Mar. Pollut. Bull. 3: 7-10.

NOTA, D. J. G., AND D.H. LORING. 1964. Recent depositional conditions in the St. Lawrence River and Gulf. A reconnaissance survey. Mar. Geol. 2: 198-235.

RANTALA, R. T. T., AND D. H. LORING. 1975. Multi-element analysis of silicate rocks and marine sediments by atomic absorption spectrophotometry. Atom. Absorp. Newsl. 14: 117-120.

SKEI, J. M., N. B. PRICE, S. E. CALVERT, AND H. HOLTEDAHL. 1972. The distribution of heavy metals in sediments of Sorfjord, West Norway. Water Air Soil Pollut. 1: 452-461.

SMITH, J.N., AND D.H. LORING. 1981. Geochronology for mercury pollution in the sediments of the Saguenay Fjord, Quebec. Environ. Sci. Tech. 15: 944-951.

STONER, J. R. 1974. Trace element geochemistry of particulate matter and water from the marine environment. Thesis, University of Liverpool, Liverpool, England. 336 p.

SUNDBY, B., AND N. SILVERBERG. 1985. Manganese fluxes in the benthic boundary layer. Limnol. Oceanogr. 30: 372-381.

WILLEY, J. D. 1976. Geochemistry and environmental implications of the surficial sediments in northern Placentia Bay, Newfoundland. Can. J. Earth Sci. 13: 1393-1410.

YEATS, P. A., B. SUNDBY, AND J. M. BEWERS. 1979. Manganese recycling in coastal waters. Mar. Chem. 8: 43-55.

CHAPTER VIII

Pollution History and Paleoclimate Signals in Sediments of the Saguenay Fjord

J. N. Smith

Marine Chemistry Division, Physical and Chemical Sciences Branch,
Department of Fisheries and Oceans, Bedford Institute of Oceanography,
P.O. Box 1006, Dartmouth, N.S. B2Y 4A2

Introduction

The Saguenay Fjord is a long (70 km), narrow (1-6 km wide), submarine valley carved out of the crystalline rocks of the Canadian Shield, draining into the St. Lawrence Estuary 180 km east of Quebec City (Fig. VIII.1). Having high rock walls extending from several hundred metres above to several hundred metres below sea level, the Saguenay Fjord represents a dramatic landscape which has long been shrouded in Indian legend and Anglo-French awe and superstition. The geographical isolation of the Saguenay, dominated by the deep chasm of the Fjord and its large (78 000 km^2) drainage basin has resulted in a unique cultural identity for this region. It is also for geographical reasons that the Saguenay Fjord has a unique environment for the study of fundamental oceanographic processes which has been exploited in a series of scientific investigations which are the focus of this chapter.

One of the important goals of marine geochemistry is the prediction of the partitioning of radionuclides and their stable analogues between different environmental phases and the rates of exchange between these phases. Measurements of these parameters are critical to the understanding of global geochemical cycles and have an important

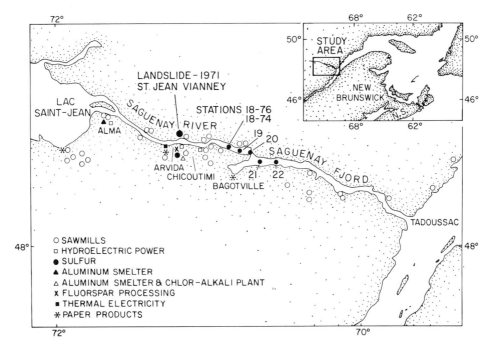

FIG. VIII.1. Location of the Saguenay Fjord in eastern Quebec (from Schafer et al. 1980, with permission).

bearing on problems associated with the impact of pollution on the oceans. The enormous complexity of many natural systems in terms of ecology, climate and geology imposes severe constraints on the design of experiments directed towards quantitative determination of transport parameters. However, difficulties encountered in these studies can, in some cases, be alleviated by studying an isolated environmental compartment and employing simplifying assumptions in modeling radionuclide transport. One of the best examples of such an isolated compartment is a drainage basin in which the physico-chemical processes governing the environmental transport of radionuclides and other substances can be characterized by studies of the time dependence of their transit through the drainage basin. The dynamics of these processes can be established either by measurements conducted over time periods appropriate to the rates of the processes themselves or by examination of a proxy record of geochemical change found in an undisturbed environmental compartment such as the sediments.

The utility of the Saguenay drainage basin as a marine geochemistry "laboratory" resulted primarily from the identification of a unique depositional regime near the head of the Fjord containing an extremely well-preserved record of environmental change during the past 500 years. The time-stratigraphy for sediment deposition was established using ^{210}Pb and other radioactive dating techniques and the results were applied to the examination of the geochronology for sediment inputs of other substances. The results of these investigations have led to advances in two directions, the fields of marine contaminant transport and paleo-climatology. In the following sections the progressive development of research studies in these fields is outlined.

Contaminant Transport

Hg Pollution of the Fjord

Interest in the sedimentology of the Saguenay Fjord was largely generated by the discovery by Loring (1975) of extremely high levels of Hg in the sediments of this region (Fig. VIII.2). Loring noted that Hg levels of the order of 1-10 $\mu g \cdot g^{-1}$ measured in the Fjord were far in excess of those levels (less than 1 $\mu g \cdot g^{-1}$) detected in the fine-grained sediments of the St. Lawrence Estuary and Gulf of St. Lawrence (see Chapter I for a description of the sedimentary regime). High levels of Hg had previously been found in both commercial fish and sediments from the Great Lakes and St. Lawrence River, most of which were derived from industrial sources (Bligh 1970). Elevated levels of Hg had also been detected in fish in the Saguenay Fjord (Tam and Armstrong 1972) and it appeared that the fish and sediment contamination had the same source. Loring's results (1975) indicated that an industrial source for anthropogenic Hg was located somewhere upstream from the Saguenay Fjord, probably in the

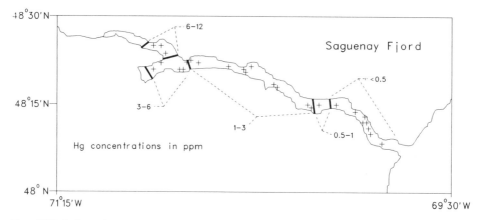

FIG. VIII.2. Distribution of total Hg in the sediments of the Saguenay Fjord (from Loring 1975, with permission).

Saguenay River. A chlor-alkali plant and several pulp and paper mills located in this region (Fig. VIII.1) were identified as the most likely sources of Hg.

Loring's (1975) results indicated that Hg concentrations were inversely related to sediment particle size with the highest concentrations detected in the finer-grained muds of the upper arm of the Fjord. Hg was positively correlated with organic carbon and with lignin, the latter being a characteristic component of terrestrial plant material. Pocklington and MacGregor (1973) had previously shown that the organic-rich sediments deposited at the head of the Saguenay Fjord were characterized by elevated C/N ratios and high levels of lignin. Pocklington had identified this region as a depositional site for terrestrial organic material, principally pulp mill wastes, with the concentration of land-derived organic matter decreasing in a seaward direction (see Chapter V). Loring's (1975) measurements of high Hg levels in sediment from this region indicated that terrestrial organic material (mainly wood fibers) was the chief carrier phase for industrially derived Hg. He concluded that Hg was either adsorbed by the organic component of the sediment or formed a complex with it and was subsequently bound to the sediment, subject to only minor return to solution due to methylation or the formation of soluble organometallic complexes.

In subsequent studies, Loring (see Chapter VII) noted that Co, Ni, Cr, V, and Cu were close to natural background levels for estuarine sediments. However, elevated levels of Pb and Zn were also detected in Saguenay Fjord sediments and these metals, inferred to have an industrial source, were associated with the same organic carrier phase with which the high levels of Hg had been associated. These industrially derived, organic wastes had been dispersed downstream from their source and deposited from suspension along with fine-grained inorganic and natural organic material. Although inputs of Pb and Zn were small compared to those of Hg, Loring (1976a) cautioned that excessive accumulation could occur in the future. In a later paper, Loring et al. (1983) reported the history of the inputs of a number of heavy metals, including Zn and Pb, into the sediments of the Saguenay, as recorded in a dated sediment core.

The extent of Hg contamination in the Saguenay Fjord was further delineated by studies carried out by Loring and Bewers (1978). A Hg inventory of 105 t was estimated for the water column and sediments in 1971, most of which was assumed to have been released from a chlor-alkali plant located at Arvida on the Saguenay River (Fig. VIII.1). The highest Hg concentrations were located in the sediments of the upper arm of the Fjord (Fig. VIII.2) where high sedimentation rates apparently led to rapid burial of the Hg rich layers. Although Hg discharges from chlor-alkali plants had been restricted by government regulations imposed in 1971, the results of Loring and Bewers (1978) indicated that high levels of Hg were still being detected in the water column of the Fjord as late as 1973. Concern for Hg contamination of the large commercial fishery in the Gulf of St. Lawrence was further heightened by the results of Bourget and Cossa (1976) which indicated that the Hg content of mussels decreased in samples collected in a seaward direction through the St. Lawrence Estuary suggesting that this concentration gradient was related to Hg discharges from the Saguenay Fjord.

Despite the wide range of Hg studies undertaken in the Saguenay Fjord and St. Lawrence Estuary and a mass-balance analysis of Hg fluxes through the system (Loring and Bewers 1978) the Hg source function remained unresolved. It was not clear whether Hg discharges from the plant had exceeded government imposed limits following the introduction of new restrictions in 1971. In addition, although other evidence indicated that the Hg was derived from the chlor-alkali plant on the Saguenay River, correlating the history of Hg inputs into the sediments with the operation of the chlor-alkali plant could provide definitive proof as to whether the chlor-alkali plant was the sole source of the Hg pollution. Clearly, alternative techniques were required to resolve these issues.

²¹⁰Pb Dating Method

Growing concern with regard to the contamination of marine and aquatic resources by anthropogenic pollutants during the 1960's and 1970's led to enhanced efforts to develop a methodology for assessing the scale and timing of pollution events. One technique for evaluating the source function for pollution of an aquatic system was to resolve the chronology for pollutant inputs to the underlying sediments. Recognition of the potential utility of this approach led to the development of new radioactive dating techniques for determining recent sedimentation rates in lake and coastal environments. The most successful of these has proven to be the ^{210}Pb dating method first applied by Koide et al. (1972) to marine systems. ^{210}Pb is a member of the naturally occurring ^{238}U decay series of radioisotopes which are among the more abundant radioactive constituents of the earth's crust. Disequilibria are generated in this decay series by the release of the inert gas ^{222}Rn from soils into the atmosphere where it rapidly decays to ^{210}Pb. Precipitation deposits ^{210}Pb on the surfaces of oceans and lakes. There it is rapidly scavenged and transported to the sediments where it appears as an excess of radioactivity, is buried and undergoes radioactive decay with a half-life of 22.3 yr. Eventually a steady state can be established in which measurements of the exponentially decreasing concentration of excess ^{210}Pb in the sediment column can be used to determine the sedimentation rate over the past 100 years of deposition. Robbins and Edgington (1975) confirmed the applicability of this naturally-occurring radioisotope as a geochronological tracer through measurements of other time-stratigraphic horizons associated with fallout radionuclides such as ^{137}Cs. This was clearly a technique that might be applied to determine the Hg pollution source function in contaminated sediments such as those of the Saguenay Fjord.

Smith and Walton (1980) used ^{210}Pb to study recent sedimentation rates in the Saguenay Fjord. They extended previous applications of this tracer by identifying a wide range of other time-stratigraphic markers based on pollen, weapons fallout tracers and microfossil assemblages which corroborated the results of ^{210}Pb dating. These studies indicated that sedimentation rates in the Saguenay Fjord decreased exponentially with increasing water depth and distance from the head of the Fjord (Fig. VIII.3) thereby identifying the Saguenay River as the principal source of suspended material deposited in the Fjord. Hg levels were also measured in sediment cores which had been dated using ^{210}Pb (Smith and Loring 1981). The initial appearance of elevated levels of Hg above background occurred at sediment depths ranging from 4 to 20 cm (Fig. VIII.4). However, the depth of Hg burial was approximately proportional to the sedimentation rate so that when a ^{210}Pb geochronology was applied to each core, the

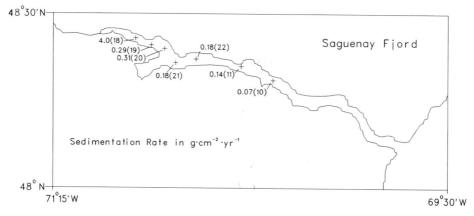

FIG. VIII.3. Sediment accumulation rates determined from ^{210}Pb dating are given for various core locations in the Fjord. Station numbers are given in parentheses (redrawn with permission from Smith and Loring 1981).

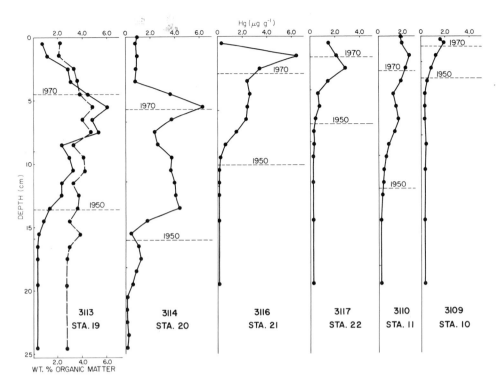

FIG. VIII.4. Hg concentration versus sediment depth for the suite of cores given in Figure VIII.3. The ^{210}Pb dates for 1950 and 1970 are provided for each core. These horizons are a function of sedimentation, density and, in some cases (core 3110), slumping (reprinted with permission from Smith and Loring 1981).

anthropogenic Hg (excess Hg above background) threshold corresponded to a depositional date of approximately 1950 (Fig. VIII. 4). A more accurate measurement of the Hg source function is the Hg sediment flux which is the product of the sedimentation rate and the Hg concentration. The largest Hg fluxes were measured at the head of the Fjord but the Hg flux exhibits the same type of geochronology at each location in the Fjord (Fig. VIII.5). The most significant finding of Smith and Loring (1981) was that the onset of anthropogenic inputs of Hg occurred in 1947 ± 3 yr according to the ^{210}Pb geochronology for the suite of cores. Since the chlor-alkali plant was constructed in 1947, the results clearly identify this plant as being the principal source of Hg contamination in the Saguenay Fjord.

The results of Smith and Loring (1981) and Schafer et al. (1980) addressed a second important point regarding Hg contamination in the Fjord. Loring and Bewers (1978) had detected high levels of Hg in fjord waters in 1973, indicating that high levels of Hg releases may have continued during the early 1970's, several years after the chlor-alkali plant had been ordered to reduce its Hg emissions. However, in each Saguenay Fjord core, the Hg flux was observed to decrease near the sediment-water interface (Fig. VIII.6). Both the timing and magnitude of the Hg fluxes near the 1971 horizon were consistent with compliance by the chlor-alkali plant with government regulations imposing restrictions on the release of Hg from these plants in their liquid effluent. Smith and Loring (1981) estimated the time-dependent source function for releases from the chlor-alkali plant and placed limits on the flushing rate of Hg from the water column of the Fjord. In a related study, Schafer et al. (1980) undertook a detailed analysis of Hg and organic matter inputs to the Fjord between 1964 and 1976 through analyses of a core collected in a well preserved depositional regime near the head of the Fjord. These results indicated that the residence time of Hg in upstream riverine sediments was of the order of 2 years and they confirmed the geochronology for Hg

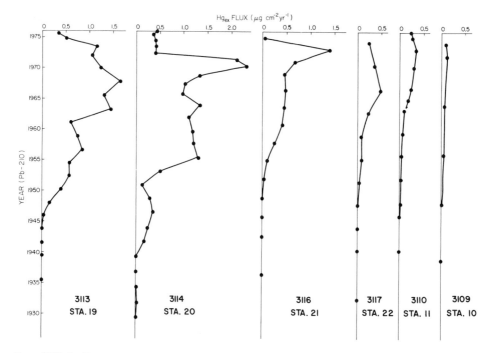

FIG. VIII.5. Excess or anthropogenic Hg (total minus background Hg) flux versus
^{210}Pb dates of deposition for each core (redrawn with permission from Smith and
Loring 1981).

inputs reported in Smith and Loring (1981). Barbeau et al. (1981a, 1981b) also
conducted studies in which the age of the oldest Hg horizon was determined by measurements of ^{137}Cs horizons associated with nuclear weapons tests. These authors
concluded that contamination of the Fjord by industrial inputs of Pb and Zn had
occurred during the post-World War II industrial development of the Saguenay region.

Pulp Wastes and PAH Contaminants

Many types of contaminants, in addition to heavy metals, have been identified in
the sediments of the Saguenay Fjord. A number of authors (Loring and Nota 1968;
Loring 1975; Pocklington and MacGregor 1973; Tan and Strain 1979a) have drawn
attention to the large quantities of terrestrial organic material deposited in the sediments
at the head of the Fjord. Ironically, it is the high concentration of organic material which
has facilitated the sedimentological studies of other pollution inputs. This is because
the organic concentrations are sufficiently great to deplete the oxygen reserves of the
sediments resulting in an anoxic benthic zone, essentially devoid of bioturbating organisms. The absence of sediment mixing by marine fauna has led to the exceptionally
good preservation of sediment stratigraphy and facilitated the precise dating of these
sediments using radioactive methods.

The historical record for the growth of the pulp and paper industry in the Saguenay
region can be resolved from an analysis of the vertical distribution of organic material
in the sediments (Schafer et al. 1980; Smith and Schafer 1987). An increase in the
organic matter content above background levels is first evident in dated sediments
deposited in 1910–12 (Fig. VIII.7). This was co-incident with the installation of the
first large pulp machines in the pulp mill at Kenogami on the Saguenay River (Fig.
VIII.1). The expansion of the Alma and Kenogami pulp mills during the 1920's is
reflected in higher sediment concentrations for this period while the bankruptcy of the
Saguenay pulp industry during the 1930's is marked by a sudden decline in the organic

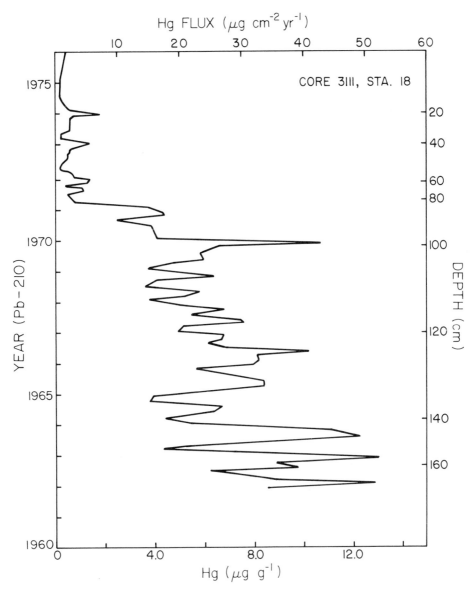

FIG. VIII.6. Hg flux as function of ^{210}Pb date of deposition for core 3111 (Station 18). Abrupt decrease in 1971 reflects decrease in Hg releases from the chlor-alkali plant (reprinted with permission from Smith and Loring 1981).

concentration profile (Smith and Schafer 1987). The organic matter flux to the sediments increased again during the post World War II period of pulp mill expansion and then declined for a second time in sediments deposited during the 1960's, probably in response to the introduction of pollution control devices in pulp mills at this time.

Recently, other types of contaminants have become major sources of environmental concern. Polyaromatic hydrocarbons (PAH), some of which are extremely carcinogenic substances, are released during most types of combustion processes and have been detected at high levels in the sediments of the Saguenay Fjord (Martel et al. 1986). Identification of the sources for these contaminants is particularly important in this heavily industrialized region because Chicoutimi has one of the highest incidences of cancer in Canada. The inferred source of the PAH releases is a large aluminum refining complex having facilities at Arvida and Alma (Fig. VIII.1) where large quantities of PAH's are released as a by-product of electrolytic aluminum refining processes

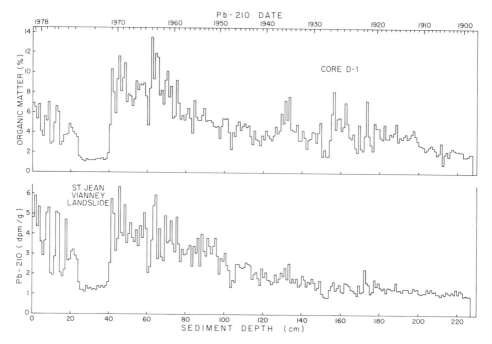

FIG. VIII.7. % organic matter and ^{210}Pb distributions for core D-1 versus sediment depth and ^{210}Pb date of deposition (upper axis) determined using a ^{210}Pb constant flux model. ^{210}Pb concentrations are in disintegrations per minute (dpm) per gram of dried sediment (from Smith and Schafer 1987, with permission).

(Martel et al. 1986). Smith and Levy (1988) have determined the detailed source function for PAH fluxes to the sediments by analyses on a precisely dated sediment core having a geochronological resolution of better than one year. Their results indicate that the PAH inventory is closely associated with the same carrier phase of terrestrial organic matter, mainly wood fibre, with which Hg is associated. The major PAH flux to the sediments of the Fjord is associated with the direct discharge into the Saguenay River of wastes recovered from atmospheric filters or "scrubbers". A smaller component of the sediment PAH flux appears to be associated with direct atmospheric releases of PAH's followed by local deposition in the Fjord. Smith and Levy (1988) also indicated that reduced inputs of PAH's into the Saguenay River since the late 1970's are due to the implementation of more efficient cleaning and filtering procedures by the refinery in 1976.

Drainage Basin Model for Radionuclide Transport

Initially, fallout radionuclides such as ^{137}Cs and ^{239,240}Pu which are derived from nuclear weapons tests conducted in the atmosphere, were measured in sediments in order to confirm ^{210}Pb sediment geochronologies. Atmospheric weapons tests began in 1952, peaked in frequency and magnitude in 1958 and again in 1962, and were largely discontinued a few years later, thereby giving a very distinct shape to the source function for fallout radioactivity to the earth's surface. However, ^{210}Pb dating in Saguenay Fjord sediments proved to be so precise that the sediment fallout profiles could be analysed to reveal subtle features associated with their geochemical cycling through aquatic and marine systems. In order to interpret fallout radionuclide inputs to sediments a two component drainage basin model was developed for the transport of particle-associated radionuclides through the soil and water components of the watershed (Smith and Ellis 1981, 1982). A comparison between the time histories for the atmospheric radionuclide inputs to the drainage basin and the appearance of these signals in Saguenay Fjord sediments was used to parameterize the transport of the radionuclides through the drainage basin. This model was used to predict ^{137}Cs and Pu sediment-depth distributions for different input functions and different drainage basin residence times which could be compared with experimental profiles to reveal details of their transport properties (Smith and Ellis 1982).

The sediment-depth distribution of [239,240]Pu for a core (D-1) collected at the head of the Fjord is illustrated in Fig. VIII.8 with the ^{210}Pb date for each sediment stratum at the top of the figure. The onset of the Pu signal in 1952 and the maxima in the Pu signal in the late 1950's and early 1960's are in agreement with the fallout record. The maxima and minima in this profile are mainly due to pulsed inputs of coarser material, deficient in Pu, during spring river discharge and landslide events and it is necessary to employ the Pu flux profile (equal to the product of the Pu concentration and sedimentation rate for each interval) in analyses of the dynamics of Pu transport through the drainage basin. The Pu flux distribution (Fig. VIII.9) is compared with values of the flux predicted by the drainage basin model for rate constant values given in Smith et al. (1987). These results indicate that the residence time for Pu in the soils of the drainage basin is of the order of 3000 yr before it is eroded in association with soil particles and transported by fluvial systems through the outlet of the system. In contrast, ^{137}Cs was observed to have a significantly reduced drainage basin residence time of 1000 yr, indicative of its greater mobility within the system (Smith et al. 1987).

The persistence of Pu in the soil component of the drainage basin was confirmed by studies of [238]Pu in sediments. In 1964, a navigational satellite powered by a

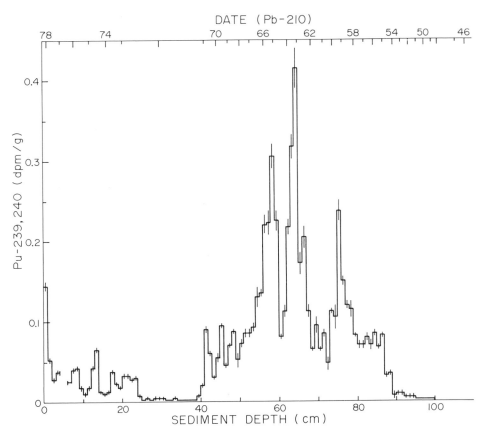

FIG. VIII.8. [239,240]Pu activity distribution versus sediment depth and ^{210}Pb date of deposition (upper axis) for core D-1 (from Smith et al. 1987, with permission).

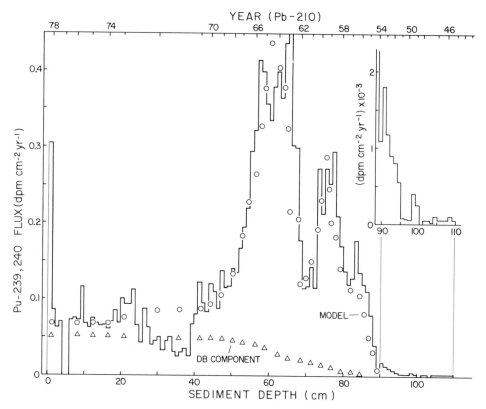

FIG. VIII.9. ^{239,240}Pu flux versus sediment depth and ^{210}Pb date of deposition for core D-1. Fluxes computed for the drainage basin model given as open circles while the flux derived exclusively by erosion from the drainage basin is indicated by open triangles. Inset illustrates Pu flux during initial weapons tests in the early 1950's (from Smith et al. 1987, with permission).

SNAP-9A nuclear power generator containing 17 kCi of ^{238}Pu burned up upon re-entry over the Indian Ocean resulting in a tripling of the global inventory of ^{238}Pu. This pulsed input of ^{238}Pu to the earth's surface resulted in an abrupt increase in the ^{238}Pu/^{239,240}Pu ratio (above the background value of 0.024 due to nuclear weapons tests) in atmospheric particulates (Thomas and Perkins 1975) for the decade following the event as illustrated in Fig. VIII.10. The ^{238}Pu/^{239,240}Pu ratio in dated Saguenay sediments also increases in response to the SNAP-9A event (Fig. VIII.10) in the late 1960's. But in this case the ^{238}Pu/^{239,240}Pu ratio increase lags the increase in the ratio in atmospheric particulates owing to delays associated with Pu transport through the soil and water components of the drainage basin. The atmospheric input function (atmospheric particulate concentration) for ^{238}Pu can be introduced into the drainage basin model using the same rate constants employed in the study of weapons fallout inputs and the model predictions are in good agreement with the observed sediment profiles (Fig. VIII.10). These studies also confirm the absence of post-depositional migration of Pu isotopes in anoxic marine sediments as any remobilization phenomena would rapidly decrease the ^{238}Pu/^{239,240}Pu concentration gradient in core D-1 by isotopic equilibration.

Paleo-Climatological Studies in the Saguenay Fjord

Drainage basins exhibit great diversity in terms of landscape, climate and ecology and each has its own hydrological characteristics which are reflected in its parti-

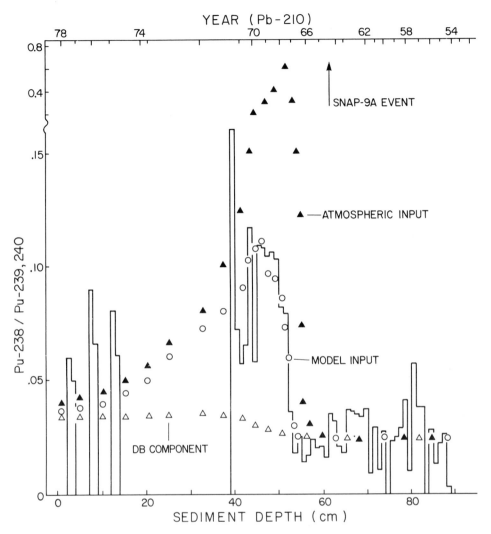

FIG. VIII.10. ^{238}Pu/^{239,240}Pu ratio in core D-1 versus sediment depth and date of deposition. Atmospheric input represents activity ratio on air particles while model predictions for sediment ratio given by open circles (from Smith et al. 1987, with permission).

cle/water discharge characteristics. River discharge through temperate watersheds is modulated primarily on a seasonal time scale, but is also affected by secondary frequency components related to longer-term climatological trends. Particle erosion and mass transport through these systems are sensitive to seasonal variations in temperature and precipitation owing to the influence of these variables on the hydrological cycle. The precise determination of seasonal variations in sedimentation rates in the anoxic, well-preserved sediments at the head of the Fjord afforded an excellent opportunity to study the influence of historical changes in climate on the depositional features at the mouth of a drainage basin.

Yeats and Bewers (1976) studied the seasonal variations in dissolved trace metal concentrations in the waters of the Saguenay Fjord and noted the enhancement in some metal concentrations during the spring freshet. Sundby and Loring (1978) examined seasonal changes in metal content of suspended particulate matter (SPM) in the Saguenay Fjord and reported a pronounced increase in the concentrations of SPM during the spring runoff, particularly evident in surface waters at the head of the Fjord.

SPM concentrations can be enhanced by an order of magnitude during the April/May spring freshet when river discharge increases by a factor of 2–3 in excess of the mean 20th century discharge rate for the Saguenay River of 1600 $m^3 \cdot s^{-1}$. The first evidence for a seasonal cycle in sedimentological properties was reported by Smith and Walton (1980) who observed seasonal variations in textural and colour properties in sediments at the head of the Fjord. The single, most pronounced feature in these cores was a gray layer of Leda clays, up to 30 cm thick in sediments near the mouth of the Saguenay River, resulting from a massive landslide near St-Jean-Vianney in May, 1971. Approximately 25 million t of Champlain Sea marine sands and clays were displaced into the Saguenay River during the heavy spring runoff, resulting in the loss of 30 lives and 40 homes. Much of this material was transported through the Saguenay River and deposited in the Fjord where it had produced a distinctive sub-surface gray layer of sediment which thins seaward from the head of the Fjord. The landslide origin of these gray sediment layers, characterized by sharply reduced organic matter concentrations in core D-1 (Fig. VIII.7), was confirmed by the presence of a distinctly marine microfossil assemblage (Schafer et al. 1980) characteristic of aged Mer de LaFlamme sediments not found at present in the upper regions of the Fjord. The gray clay landslide horizon provides a May, 1971 time-stratigraphic horizon as a calibration isochron for all sediment geochronological studies in Saguenay sediments.

Landslide-displaced sediment, initially deposited in upstream Saguenay River basins, was resuspended and deposited in the Fjord during freshets of subsequent years to produce successive gray clay layers of sediment overlying the 1971 landslide horizon (Smith and Walton 1980). Subsequent detailed studies of ^{210}Pb distributions indicated that sediment resuspension and enhanced particle transport into the Fjord during periods of high runoff had been a major mass transport mechanism over the past few hundred years and one which could be used to advantage to gain information of climatological importance. The sediment-depth distribution of ^{210}Pb in core D-1, collected near the head of the Fjord where the sedimentation rate is 2.16 $g \cdot cm^{-2} \cdot yr^{-1}$, exhibited an annually modulated sequence for sediment deposition since 1900 (Fig. VIII.7). Reduced ^{210}Pb activities are associated with coarser-grained silts and sands pulsed into the Fjord during high river discharge events which alternate with sequences of finer-grained, organic rich material having elevated ^{210}Pb activities which are deposited from suspension under conditions of reduced river discharge (Smith and Ellis 1982). The good agreement between the % organic matter and ^{210}Pb distributions suggests that the same, seasonally modulated sediment deposition mechanism governs the removal of both substances to the sediments. A constant flux ^{210}Pb model was applied to these results in order to derive the sedimentation rates for 1 cm intervals of the sediment column. Seasonal increases in the sedimentation rate associated with the spring discharge can be correlated with the historical record of river discharge for the Saguenay to provide a second geochronology for core D-1 which is in good agreement with the ^{210}Pb constant flux geochronology (Fig. VIII.11, Smith and Schafer 1987). The largest sedimentation rate events are associated both with the St-Jean-Vianney landslide, denoted by a 15 cm thick interval of gray clays deposited at sedimentation rates in excess of 15 $g \cdot cm^{-2} \cdot yr^{-1}$, and subsequent spring river discharges in 1973, 1974, and 1976. The largest June/July river discharge ever recorded in the Saguenay River, in 1928, is correlated with an unusually high sedimentation rate of 17.6 $g \cdot cm^{-2} \cdot yr^{-1}$. High spring river discharge events in 1942, 1944, and 1964 are also reflected by high sedimentation rate events for those years (Smith and Schafer 1987).

Grain size analyses of sediment intervals from core D-1 reveal a linear correlation between % sand and the sedimentation rate (Smith and Schafer 1987) for sediment deposition during the 20th century. This correlation is consistent with the mechanism outlined above whereby high sedimentation rates are associated with high river discharge events during which coarser-grained material is pulsed into the Fjord. However, a parameter having greater climatological significance is the sand flux ($g \cdot cm^{-2} \cdot yr^{-1}$) which is equal to the product of the % sand and sedimentation rate for each sediment

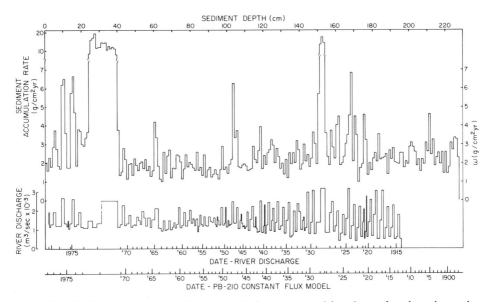

FIG. VIII.11. The quarter-monthly river discharge record has been fitted to the sediment accumulation rate distribution for core D-1 to illustrate linkage between sedimentological parameters and river discharge (from Smith and Schafer 1987, with permission).

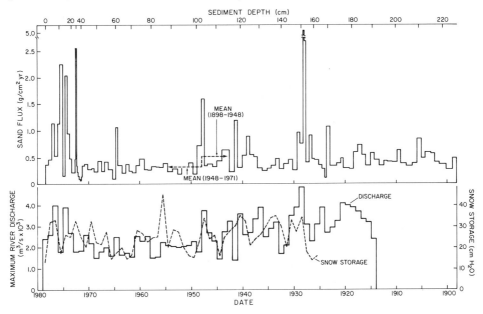

FIG. VIII.12. Sand flux versus date of deposition. The maximum mean monthly discharge level and annual snow storage function for each year are illustrated in bottom figure. Decrease in the mean sand flux in 1948 is related to the overall 20th century decrease both in snow storage and in maximum river discharge (from Smith and Schafer 1987, with permission).

interval. The correlation between sand flux and maximum river discharge is illustrated on a linear time scale for core D-1 in Fig. VIII.12. Most of the major 20th century river discharge events are reflected in the sediments by synchronous sand flux events with the 1928, 1940, 1942, 1947, and 1964 river discharge and 1971 landslide peaks being the most prominent. River discharge is related to snow storage (water equivalent

135

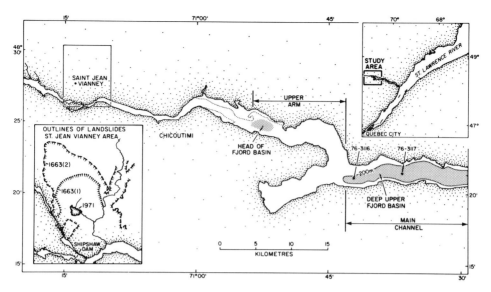

FIG. VIII.13. Outlines of landslides in the St-Jean-Vianney area in 1663 and 1971 (from Schafer and Smith 1987, with permission).

of snow stored over the previous winter), both of which exhibit a 20th century decrease with time in agreement with the corresponding decrease observed in the sand flux with time in core D-1 (Smith and Schafer 1987). Recognition of the relationship between sedimentological parameters (sand flux) and climatological parameters (snow storage, itself a function of temperature and precipitation) provides a methodology for reconstructing historical changes in river discharge and climate through analyses of the pre-20th century portions of sediment cores. Efforts are being undertaken at present to reconstruct the detailed record for Saguenay River discharge for the 19th century and to model these results to determine temperature/precipitation changes during this period.

The interpretation of pre-20th century portions of Saguenay Fjord sediment cores was assisted by the identification of time-stratigraphic horizons associated with an earthquake-triggered landslide event in 1663. This slide occurred in the present day scarp in which the 1971 St-Jean-Vianney landslide took place, although the earlier slide displaced an order of magnitude more material (Fig. VIII.13). The slide (the largest ever recorded in eastern Canada) took place shortly after one of the largest earthquakes ever recorded in Quebec, which had its epicenter in the St. Lawrence River. The landslide horizon is recorded in sediment cores as a sand horizon underlying a series of clay layers characterized by marine microfossil assemblages typical of Champlain Sea sediments (Smith and Schafer 1987). Seismic studies in the Saguenay Fjord reveal that the 1663 earthquake event may have resulted in the collapse of a vast section of the Fjord and produced a massive debris flow along the length of the system. Studies of the geomorphological implications of these events are continuing to reveal insight into the mechanisms governing large submarine slides.

Summary

The Saguenay Fjord has provided an oceanographic laboratory for studies pertaining to contaminant transport through drainage basins and for investigations focusing on the North American paleoclimatological regime of the past 500 years. The key pollution issue in the Saguenay Fjord was that of establishing the source function for Hg pollution. This problem was solved by determining the geochronology for Hg inputs to the sediments using the ^{210}Pb dating method. The results of this investigation indicated that a chlor-alkali plant was, as suspected, the principal source of Hg pollution in the system. But the results also indicated that, contrary to expectations, the chlor-alkali plant had fully complied with government regulations governing its Hg releases in liquid discharges. Other contaminant transport processes which have been investigated in the Fjord include those associated with organic wastes from pulp and paper mills, with PAH releases from aluminum refining complexes and with radioactivity inputs from atmospheric nuclear weapons tests, satellite burn-ups and nuclear reactor accidents. This suite of interrelated investigations has led to the development of a general model which can be used to predict the transport parameters of anthropogenic substances released into the environment as they pass through various components of a drainage basin.

Paleoclimatological studies undertaken in the Fjord were based on experimental observations of a seasonally modulated, ^{210}Pb signal in the anoxic, unbioturbated sediments at the head of the Fjord. These results were used to determine annual variations in sedimentation rates and textural properties of the sediments whose variability was then established as a function of the magnitude of the spring river discharge. Having determined these relationships for 20th century sediments and known river discharge records, studies were undertaken on older sediments in order to reconstruct river discharge and paleoclimatological information from the 16th-19th centuries. These paleoclimatological investigations led to an analysis of the frequency and magnitude of landslide events in the Saguenay drainage basin and to studies of the relationship between earthquake events and geomorphological changes within this region. Therefore, the sedimentological focus of work in this region has shifted from sediment deposition during the past 30 years to geological and paleoclimatological events which occurred over 300 years ago.

References

BARBEAU, C., R. BOUGIÉ, AND J.-E. COTÉ. 1981a. Temporal and spatial variations of mercury, lead, zinc, and copper in sediments of the Saguenay Fjord. Can. J. Earth Sci. 18: 1065-1074.

　　　　1981b. Variations spatiales et temporelles du cesium-137 et du carbone dans des sédiments du Fjord du Saguenay. Can. J. Earth Sci. 18: 1004-1011.

BLIGH, E. G. 1970. Mercury and the contamination of freshwater fish. Fish. Res. Board Can. MS. Rep. 1088: 27 p.

BOURGET, E., AND D. COSSA. 1976. Mercury content of mussels from the St. Lawrence Estuary and Northwestern Gulf of St. Lawrence. Mar. Pollut. Bull. 7: 237-239.

KOIDE, M., K. W. BRULAND, AND E. D. GOLDBERG. 1972. Th-228/Th-232 and Pb-210 geochronologies in marine and lake sediments. Geochim. Cosmochi. Acta 37: 1171-1187.

LORING, D. H. 1975. Mercury in the sediments of the Gulf of St. Lawrence. Can. J. Earth Sci. 12: 1219-1237.

　　　　1976a. The distribution and partition of zinc, copper, and lead in the sediments of the Saguenay Fjord. Can. J. Earth Sci. 13: 960-971.

LORING, D. H. AND J. M. BEWERS. 1978. Geochemical mass balances for mercury in a Canadian Fjord. Chem. Geol. 22: 309-330.

LORING, D. H. AND D. J. G. NOTA. 1968. Occurrence and significance of iron, manganese, and titanium in glacial marine sediments from the estuary of the St. Lawrence river. J. Fish. Res. Board Can. 25: 2327-2347.

LORING, D. H., R. T. T. RANTALA, AND J. N. SMITH. 1983. Response time of Saguenay Fjord sediments to metal contamination. Environ. Biogeochem. Ecol. Bull. 35: 59-72.

MARTEL, L., M. J. GAGNON, R. MASSE, A. LECLERC, AND L. TREMBLAY. 1986. Polycyclic aromatic hydrocarbons in sediments from the Saguenay Fjord, Canada. Bull. Environ. Contam. Toxicol. 37: 133-140.

POCKLINGTON, R., AND C. D. MACGREGOR. 1973. The determination of lignin in marine sediments and particulate form in seawater. Int. J. Environ. Anal. Chem. 3: 81-93.

ROBBINS, J. A., AND D. N. EDGINGTON. 1975. Determination of recent sedimentation rates in Lake Michigan using Pb-210 and Cs-137. Geochim. Cosmochim. Acta 39: 285-304.

SCHAFER, C. T., AND J. N. SMITH. 1987. Hypothesis for a submarine landslide and cohesionless sediment flows resulting from a 17th century earthquake-triggered landslide in Quebec, Canada. Geo-Marine Lett. 7: 31-37.

SCHAFER, C. T., J. N. SMITH, AND D. H. LORING. 1980. Recent sedimentation events at the head of the Saguenay Fjord, Canada. Environ. Geol. 3: 139-150.

SMITH, J. N., AND K. M. ELLIS. 1981. Transport mechanism for fallout ^{137}Cs to estuarine sediments, p. 119-130. In Impacts of radionuclide releases into the marine environment. International Atomic Energy Agency, Vienna.

    1982. Transport mechanism for Pb-210, Cs-137, and Pu fallout radionuclides through fluvial-marine systems. Geochim. Cosmochim. Acta 46: 941-954.

SMITH, J. N., AND E. LEVY. 1988. A geochronology for PAH contamination recorded in the sediments of the Saguenay Fjord. Unpubl. manuscript.

SMITH, J. N., AND D. H. LORING. 1981. Geochronology for mercury pollution in the sediments of the Saguenay Fjord, Quebec. Environ. Sci. Tech. 15: 944-951.

SMITH, J. N., AND C. T. SCHAFER. 1987. A 20th century record of climatologically modulated sediment accumulation in a Canadian Fjord. Quat. Res. 27: 232-247.

SMITH, J. N., AND A. WALTON. 1980. Sediment accumulation rates and geochronologies measured in the Saguenay Fjord using the Pb-210 dating method. Geochim. Cosmochim. Acta 44: 225-240.

SMITH, J. N., K. M. ELLIS, AND D. M. NELSON. 1987. Time-dependent modeling of fallout radionuclide transport in a drainage basin: Significance of "slow" erosional and "fast" hydrological components. Chem. Geol. 63: 157-180.

SUNDBY, B., AND D. H. LORING. 1978. Geochemistry of suspended particulate matter in the Saguenay Fjord. Can. J. Earth Sci. 15: 1002-1011.

TAM, K.C., AND F.A.J. ARMSTRONG. 1972. Mercury contamination in fish from Canadian waters. p. 4-21. In J. F. Uthe [ed.] Mercury in the aquatic environment: A summary of research carried out by the Freshwater Institute 1970-1971. Fish. Res. Board Can. MS Rep. 1167.

TAN, F. C., AND P. M. STRAIN. 1979a. Carbon isotope ratios of particulate organic matter in the Gulf of St. Lawrence. J. Fish. Res. Board Can. 36: 678-682.

THOMAS, C. W., AND R. W. PERKINS. 1975. Transuranium elements in the atmosphere. Health Saf. Lab. Environ. Q. 291: 1-80.

YEATS, P. A., AND J. M. BEWERS. 1976. Trace metals in the waters of the Saguenay Fjord. Can. J. Earth Sci. 13: 1319-1327.

CHAPTER IX

Petroleum Residues in the Waters of the Gulf of St. Lawrence[1]

Eric M. Levy

Marine Chemistry Division, Physical and Chemical Sciences Branch,
Department of Fisheries and Oceans, Bedford Institute of Oceanography,
P.O. Box 1006, Dartmouth, N.S. B2Y 4A2

Introduction

The Gulf of St. Lawrence is the most important of the Canadian marginal seas in terms of its economic and social significance to Canada, a status that can only be maintained if sound management of the region is practiced. That is, it will only be possible to maintain the harvest of its living resources at economically viable levels while continuing the traditional use of the Gulf as a corridor for the transportation of industrial raw materials and manufactured products (including crude and refined oil in bulk) to and from the interior of the continent, and to accommodate new uses of the region such as exploration for potential hydrocarbon reserves, if the appropriate measures are taken to prevent the deterioration of its marine environment. Essential to sound environmental management of the region is reliable information concerning its chemical, physical, and biological oceanography. One facet of this is the incidence and distribution of potentially harmful substances, including oil. Accordingly, two preliminary surveys of the incidence and distribution of petroleum residues were carried out in 1971 and 1972. This was followed by a program to measure background levels of these substances in the various sectors of the Gulf and to determine how these levels change over time. Between 1971 and 1979, ten surveys of petroleum residues in the Gulf were carried out by the Chemical Oceanography Division at the Bedford Institute of Oceanography. This chapter summarizes the results and presents a general overview of the distribution of oil-related contamination in the Gulf and how these levels changed over this decade.

Sampling and Analytical Methods

To ensure that the data collected during this program would not suffer from the ambiguities that would inevitably arise from changes in sampling and analytical techniques, the following procedures were used throughout.

Floating Particulate Forms of Petroleum Residues

Weathered particulate and relatively unweathered semi-liquid oil floating on the sea surface were collected by towing a modified version of a neuston sampler (Levy and Walton 1971). The sampler consisted of a 243 μm mesh nylon plankton net attached to an open-ended aluminum box. The sampler was rigged so that it was towed to the side of the ship, thereby passing through a portion of the sea surface that was not disturbed by the passage of the vessel. Towing for a distance of 1 nautical mile (1.9 km) at a speed of 5-6 knots sampled about 740 m^2 of sea surface to a depth of $\approx$15-20 cm. On retrieval, the contents of the net were washed into clean jars and frozen for subsequent analysis. When a large amount of zooplankton, fragments of sea weed, or other floating debris was collected, this was removed by hand before bottling the sample. Any fresh oil adhering to the net was removed by extraction with carbon tetrachloride.

---

[1] Most of this article was published previously as: Levy, E. M. 1985. Background levels of dissolved/dispersed petroleum residues in the Gulf of St. Lawrence 1970-1979. Can. J. Fish. Aquat. Sci. 42: 544-555.

On return to the laboratory, any oil in the sample was quantified gravimetrically. The samples were thawed, and if sufficient were present, the oil particles were hand-picked from plankton or other particulate material, dried, and weighed. Otherwise, the oil was extracted from the sample with carbon tetrachloride, and the residue remaining after evaporation of the solvent was weighed.

Dissolved / Dispersed Petroleum Residues in the Water Column

Samples were collected throughout the water column, usually at standard oceanographic depths, using Niskin samplers on a hydrographic wire or in a CTD-Rosette assembly. Subsamples (1 L) were immediately extracted twice with redistilled carbon tetrachloride and the extracts were stored in the dark until analysis. The petroleum residues were transferred to hexane, and the intensity of fluorescence at 360 nm on excitation at 310 nm was compared with the responses from a series of solutions containing known amounts of Bunker C fuel oil (the intensity of the fluorescence emission of this oil is 0.221 times that of chrysene — Levy 1977). These procedures are capable of providing reliable data when carried out as described and reasonable care is taken to avoid contamination from the ship and other sources during processing and analysis (Levy 1979a, 1985).

Dissolved / Dispersed Petroleum Residues in the Sea Surface Microlayer

Samples of water from the sea surface microlayer were collected during the last cruise of the program by multiple dips of a modified version of the Garrett screen sampler (Garrett 1965). The samples were analyzed as described for samples of water from the water column.

Sampling Program

Because other demands on ship time precluded maintaining a regular sampling schedule, most of the studies of petroleum residues in the Gulf of St. Lawrence were on cruises of opportunity. It was not possible to carry out a detailed study of the entire region during every cruise. Heavy ice prevented sampling in some areas during several cruises. Consequently, the density of sampling and the locations of the sampling sites differed from one cruise to another. Nevertheless, it is possible to calculate background levels for the various sectors of the Gulf and to estimate the extent of the temporal change that has occurred. Sampling locations for each survey are described below (for further detail on data available from these cruises, see the Appendix).

Data Analysis

During this program, data from more than 2000 samples were obtained at 448 stations occupied during ten oceanographic cruises conducted between 1971 and 1979. From these data, a general model of the incidence and distribution of petroleum residues in the Gulf of St. Lawrence was derived and from this model changes in the background levels that occurred during the decade became evident (Levy 1985). To facilitate the analysis and interpretation of this large data set, the Gulf region was subdivided into eleven zones (Fig. IX.1): zone A is seaward of Cabot Strait; zone B is Cabot Strait; zone C is the central portion of the Gulf and is dominated by the deep inflowing water in the Laurentian Channel; zone D is the northeastern Gulf and includes any exchange through the Strait of Belle Isle with the Atlantic; zone E is the Magdalen Shallows; zone F is the Jacques Cartier Passage which is the route usually taken by ships entering or leaving the Gulf through the Strait of Belle Isle; zone G includes a relatively permanent counter-clockwise gyre in the northwest and the discharge from the St. Lawrence River system; zone H and I are the lower and upper

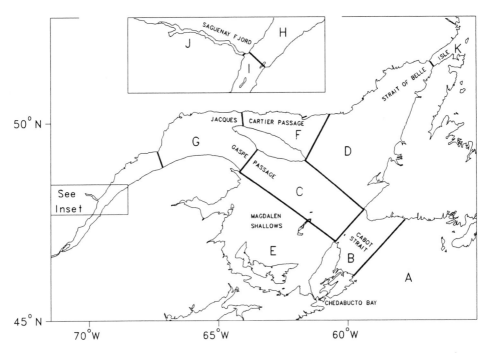

FIG. IX.1. An 11-compartment model of the Gulf of St. Lawrence (redrawn from Levy 1985).

sections of the St. Lawrence Estuary; zone J is the Saguenay Fjord; and zone K is the area of the North Atlantic outside the Strait of Belle Isle.

As was the case with the data for background levels of dissolved/dispersed petroleum residues in other regions off the east coast of Canada (Levy 1979b, 1981, 1983) and in other regions of the world ocean (Levy 1984; Levy et al. 1981), the frequency distribution histograms for the data from virtually all the cruises in the Gulf of St. Lawrence were highly skewed with most of the values near the lower end of the range of concentrations and comparatively few at the higher. For example, the histogram for the data from cruise 74-028 (Fig. IX.2) indicated that almost 60 % of the 530 samples contained less than 2 $\mu g \cdot L^{-1}$ with only a few of the remaining samples having concentrations greater than 5 $\mu g \cdot L^{-1}$, although a single value of 13.5 $\mu g \cdot L^{-1}$ was observed. Such distributions are best analyzed after logarithmic transformation of the data (Aitchison and Brown 1969) and, as illustrated by this example, the distributions for the logtransformed data were sufficiently close to normal, in most cases, to pass the χ^2-test at the 95 % confidence level. (In a few instances, the transformed data failed to satisfy the requirements of the χ^2-test at the 95 % confidence level but passed at a slightly lower level.) Consequently, the geometric mean is the appropriate measure of central tendency for the data and, as such, provides a good estimate of the general background level existent in the region at the time. After calculating the background level for each zone, t-tests were used to evaluate the significance of any differences between adjacent zones. Where these were not statistically significant, the data were combined and the analysis was repeated. Eventually, this procedure reduced the original 11-compartment model of the Gulf to a much simpler one in which the differences in the background levels of dissolved/dispersed petroleum residues between adjacent zones were statistically significant.

Since the waters of the Gulf of St. Lawrence during summer consist of three relatively distinct layers (see Chapter I), the possibility of this structure being reflected in the distribution of the dissolved/dispersed petroleum residues was examined by comparing the geometric means for 50 m intervals of depth through the water column and

141

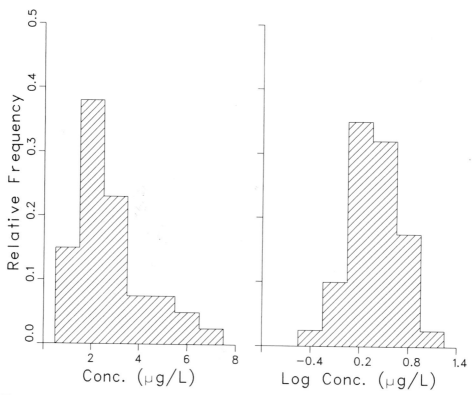

FIG. IX.2. Frequency distribution histograms of the untransformed (left) and the log-transformed (right) concentration data from cruise 74-028 (redrawn from Levy 1985).

by comparing the geometric means for the upper 100, 200, and 300 m with those for the remainder of the water column. In neither case, however, could a clear relationship between concentration and depth be established. Similarly, no direct relationship between the concentration of petroleum residues and temperature, salinity, dissolved oxygen, or nutrient concentration (all of which reflect the vertical structure) was found. Thus, the geographical distribution of dissolved/dispersed petroleum residues in the Gulf of St. Lawrence is not related simply to natural biological or geochemical processes and may be adequately described in terms of a model based solely on oceanographic/geographic factors.

Results

Cruise 70-026 (15-22 July 1970)

The first survey in the Gulf of St. Lawrence was an adjunct to an assessment of the impact of a Bunker C spill in 1970 from the tanker *Arrow* on the background level of dissolved/dispersed petroleum residues in the waters of the Scotian Shelf (Levy 1971a) and the rate at which the pre-spill levels were restored (Levy 1972). The main objective of the Gulf survey was to collect comparative data for a region having oceanographic conditions similar to those of the Scotian Shelf but one which could not have been influenced by the *Arrow* spill. A secondary objective was to provide data along a line of stations through the Gulf to test the hypothesis that the major input of petroleum-related substances to the Gulf occurs by way of the St. Lawrence River; that is, oil associated with the run-off from the St. Lawrence drainage basin, the convergence of cargo ship and tanker traffic in the upper Estuary, and losses from the oil refineries in Montreal (Levy 1971a).

Samples collected near the surface (1 m), at mid-depth, and just above the bottom had concentrations ranging from 1.3 to 5.5 $\mu g \cdot L^{-1}$. While only thirteen stations were sampled, this survey identified several intriguing features. First, there seemed to be a relatively low and uniform level of dissolved/dispersed petroleum residues present throughout the region with no obvious relationship between the concentration and the location or depth at which the samples were collected. Second, despite the run-off from the St. Lawrence drainage basin and the convergence of ship traffic, the anticipated increase in concentration in proceeding through the lower and upper Estuary into the River was not observed. To explain its absence, it was postulated that natural chemical and biological processes of degradation, adsorption onto settling suspended particles, and other removal processes were sufficiently effective to negate the assumed inputs.

Although there were too few data for a more detailed analysis, this cruise demonstrated that the incidence and distribution of dissolved/dispersed petroleum residues in the Gulf of St. Lawrence was not as expected and indicated that further study was required.

Cruise 71-027 (5 July - 8 August 1971)

Because of the intriguing results of the previous year's cruise, a more comprehensive survey was carried out in 1971 during which concentrations of both floating particulate and dissolved/dispersed petroleum residues in the Gulf were measured (Levy and Walton 1973).

No indication of floating petroleum residues was found in the samples from 20 of the sites studied, whereas relatively fresh oil adhered to the net at nine locations and weathered residues were found at 23. The most obvious feature of the data was the high values at two stations on the Magdalen Shelf in the vicinity ($47°25.0'$N, $63°19.0'$W) of where the barge, *Irving Whale*, sank in 1970 with its cargo of Bunker C fuel oil. At the time of sampling, oil was present on the sea surface as pancake-like patches and rope-like chunks and as a slick of less than 1 km long and 0.5 km wide (Levy 1971b). Since the slick persisted despite the action of winds and surface currents, it was evident that the surface contamination was being replenished by escape of oil from the sunken hull.

The distribution pattern of floating forms of oil elsewhere in the Gulf and the fact that the higher concentrations were invariably in fresh unweathered condition suggested that discharges from ships were responsible. Concentrations were generally less than 50 $\mu g / m^2$, with the lowest values in the St. Lawrence Estuary and over the Magdalen Shallows, well away from the leakage from the *Irving Whale*.

Concentrations of dissolved/dispersed petroleum residues near the surface, at mid-depth, and near the bottom within the Gulf were generally less than 5 $\mu g \cdot L^{-1}$ in the southern and western regions and between 5 and 10 $\mu g \cdot L^{-1}$ in the north and east, whereas concentrations in Cabot Strait, and in the surface water of the Atlantic just beyond, were as high as 18 $\mu g \cdot L^{-1}$ (Levy and Walton 1973). Since concentrations in the Gulf were lower than those in the surface water entering the Gulf, it was hypothesized that these inflowing waters were the major source of dissolved/dispersed petroleum residues in the Gulf. On this basis, the observed distribution of these substances in the Gulf was readily explained in terms of the general circulation of water in the Gulf and further supported the contention that inputs of petroleum residues from land drainage, domestic and industrial wastes, and discharges from shipping were not the major sources of petroleum residues in the Gulf of St. Lawrence at this time. Rather, the dominant source appeared to be associated with the surface water that enters the Gulf through Cabot Strait.

These two initial surveys demonstrated the need for a better understanding of the incidence and distribution of petroleum residues in the Gulf of St. Lawrence, particularly in view of the projected increases in shipping, the possibility of increased offshore oil exploration, and the importance of fisheries protection. In addition, they served as a convenient starting point for a program to measure background levels of dis-

TABLE IX.1. Background levels of dissolved/dispersed petroleum residues in the Gulf of St. Lawrence. G.M. is the geometric mean concentration, and Distribution range gives the concentrations corresponding to $\pm$ 1σ of the logtransformed data (all in $\mu g \cdot L^{-1}$). July/August, 1971 (Cruise 71-027).

Region	G.M.	Distribution range	n
A. Cabot Strait approaches	6.32	1.06 – 37.7	3
B. Cabot Strait	5.96	3.39 – 10.5	21
C. Central Gulf	0.85	0.45 – 1.61	6
D. Northeast Gulf	3.55	1.84 – 6.89	51
E. Magdalen Shallows	1.01	0.52 – 1.97	87
F. Jacques Cartier Passage	3.35	1.97 – 5.70	20
G. Gaspe Passage	1.32	0.58 – 2.98	36
H. Lower Estuary	0.95	0.39 – 2.35	18
J. Saguenay Fjord	2.75	1.61 – 4.74	15
K. Belle Isle approaches	—	—	—
Overall data set	1.82	0.71 – 4.62	257

solved/dispersed petroleum residues over a sufficient period to reveal longer-term trends. Although more detailed interpretation, in terms of background levels over broad geographical areas, was not considered possible because the sampling had been limited to surface, mid-depth, and near-bottom, detailed analysis of the data from subsequent cruises again failed to demonstrate any clear relationships between depth and the concentration of dissolved/dispersed petroleum residues in the Gulf. Consequently, it is valid to calculate baseline levels from these data and to include them in comparisons with the levels calculated from data obtained at standard oceanographic depths on subsequent cruises.

Concentrations of dissolved/dispersed petroleum residues in the water column at this time ranged from 0.5 to 10 $\mu g \cdot L^{-1}$ in the Gulf and from 4 to 18 $\mu g \cdot L^{-1}$ in the seaward approaches to Cabot Strait. Although there were too few values in most of the sectors to establish their statistical distribution, the geometric means indicated that the lowest levels were in the lower Estuary and the southwestern part of the Gulf (Table IX.1). After combining the data for those zones in which the apparent differences between geometric means were not significant, a 4-component model emerged comprising Cabot Strait and the area seaward, the northeastern region, and the entire southwestern region including the St. Lawrence Estuary, and the Saguenay Fjord where the background levels were 6.0, 3.5, 1.1, and 2.1 $\mu g \cdot L^{-1}$ respectively (Fig. IX.3a). This description, therefore, illustrates the conclusion drawn earlier that the major source of dissolved/dispersed petroleum residues in the Gulf was Atlantic water that entered through Cabot Strait and was not discharge from the St. Lawrence and Saguenay Rivers (Levy and Walton 1973).

In addition to the general survey, a detailed study of dissolved/dispersed petroleum residues in the water column was carried out around the *Irving Whale* site. Although unweathered oil and a slick were present on the sea surface, the level of dissolved/dispersed petroleum residues in the water column at this location (1.2 $\mu g \cdot L^{-1}$) was not significantly higher than that elsewhere on the Magdalen Shallows. Thus, while the oil on the sea surface indicated that oil continued to leak from the sunken barge, this was not having a detectable impact on the background level of dissolved/dispersed petroleum residues in the underlying water column.

Finally, to examine the role of potential land-based inputs of petroleum residues to the Gulf, samples were collected from the Saguenay Fjord and Chaleur Bay. Since the background level in the Saguenay Fjord was higher than that for the adjoining St. Lawrence estuary, it seems that there was a sufficiently large local input of aromatic substances (presumably PAH's derived from industries in the area — see Chapter I) to have a significant impact on levels in the Saguenay Fjord. It does not appear, however, that these constitute a significant input to the Gulf in general. In Chaleur Bay,

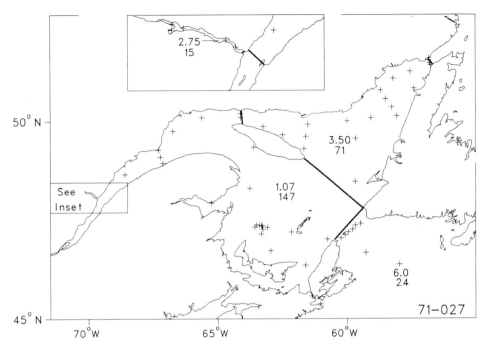

FIG. IX.3a. Background levels of dissolved/dispersed petroleum residues (μg/L — upper number) and number of samples (lower number) in different sectors of the Gulf of St. Lawrence. +'s = sample locations; heavy lines separate subregions with statistically different averages. Redrawn from Levy 1985).

on the other hand, the background level at the time (0.67 μg·L^{-1}) was well below that elsewhere on the Magdalen Shallows and, therefore, oil-containing wastes from the forest and mining industries in this area were not a significant source of oil contamination in the Gulf.

Cruise 72-017 (30 May - 13 June 1972)

Sampling during this cruise was focussed on the seaward approaches to Cabot Strait and the northeastern and Gaspe regions. Concentrations of dissolved/dispersed petroleum residues ranged from 0.6 to 15 μg·L^{-1} in the approaches to Cabot Strait and from 0.5 to 12 μg·L^{-1} in the Gulf with most of the values between 3 and 4 μg·L^{-1}. The overall data set, as well as the sub-sets for the geographical sub-areas, passed the χ^2-test after logarithmic transformation and, therefore, the geometric means provide a measure of the background levels in the various regions (Table IX.2). Combining the data for adjoining areas for which the differences between geometric means were not significant yielded a 5-compartment model (Fig. IX.3b), all sectors having background levels of 2.3 to 4.6 μg·L^{-1}. The background in the Jacques Cartier Passage was slightly lower and, as found in the previous studies, that in Cabot Strait was higher than elsewhere in the Gulf. The uniformity observed during this cruise was presumably a consequence of ambient oceanographic conditions as the cruise was carried out in early spring shortly after the Gulf became free of ice and before appreciable seasonal warming had taken place. Insufficient data were available to estimate the background levels in the upper Estuary or Saguenay Fjord or to compare the level in the northeastern portion of the Gulf with that in the Atlantic outside the Strait of Belle Isle.

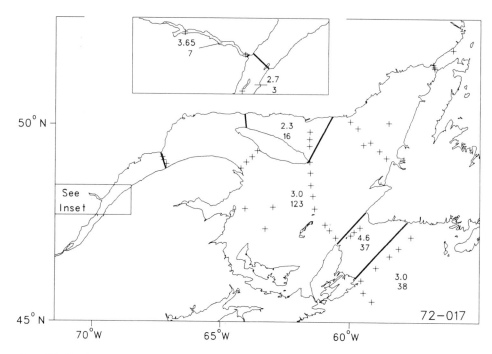

FIG. IX.3b. (*Continued.*)

TABLE IX.2. Background levels of dissolved/dispersed petroleum residues in the Gulf of
St. Lawrence. G.M. is the geometric mean concentration, and Distribution
range gives the concentrations corresponding to ± 1σ of the logtransformed
data (all in $\mu g \cdot L^{-1}$). May/June 1972 (Cruise 72-017).

Region	G.M.	Distribution range	n
A. Cabot Strait approaches	3.01	1.61 - 5.61	38
B. Cabot Strait	4.59	2.55 - 8.22	37
C. Central Gulf	2.57	1.26 - 5.26	35
D. Northeast Gulf	3.21	2.25 - 4.57	43
E. Magdalen Shallows	3.11	1.77 - 5.48	10
F. Jacques Cartier Passage	2.29	1.50 - 3.44	16
G. Gaspe Passage	3.21	2.10 - 4.90	35
I. Upper Estuary	2.73	2.52 - 2.95	3
J. Saguenay Fjord	3.65	2.89 - 4.62	7
K. Belle Isle approaches	2.25	1.82 - 2.81	3
Overall data set	3.15	1.81 - 5.51	227

Cruise 73-004 (12 Feb. - 8 March 1973)

This cruise was carried out to survey as much of the Gulf as possible under conditions
of late winter, although heavy ice precluded sampling on the Magdalen Shallows and
over most of the central region.

Concentrations of dissolved/dispersed petroleum residues at this time ranged from
0.25 to 4.6 $\mu g \cdot L^{-1}$ with most of the values around 1 $\mu g \cdot L^{-1}$. While there were too
few samples in most regions to establish the statistical distribution of the data, the geometric means indicate that the background levels were relatively uniform throughout
the region with perhaps slightly higher levels in Cabot Strait and in the northeastern

146

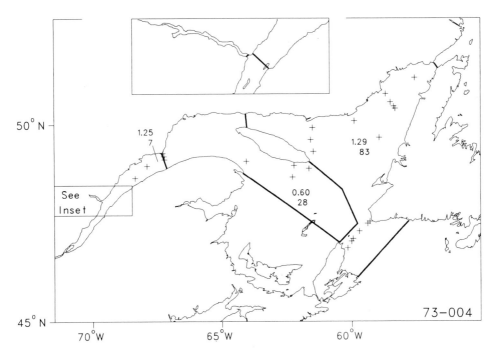

FIG. IX.3c. (*Continued.*)

TABLE IX.3. Background levels of dissolved/dispersed petroleum residues in the Gulf of St. Lawrence. G.M. is the geometric mean concentration, and Distribution range gives the concentrations corresponding to ± 1σ of the logtransformed data (all in μg·L^{-1}). Feb/March 1973 (Cruise 73-004).

Region	G.M.	Distribution range	n
B. Cabot Strait	1.37	0.80 - 2.36	33
C. Central Gulf	0.47	0.23 - 0.94	10
D. Northeast Gulf	1.29	0.70 - 2.32	42
F. Jacques Cartier Passage	1.09	0.56 - 2.11	8
G. Gaspe Passage	0.69	0.38 - 1.22	18
H. Lower Estuary	1.25	0.70 - 2.26	7
Overall data set	1.09	0.55 - 2.13	118

regions than elsewhere (Table IX.3). In this case, the distribution of dissolved/dispersed petroleum residues could be described by a 3-compartment model comprising the northeastern, central, and estuary regions with background levels of 1.3, 0.60, and 1.3 μg·L^{-1} respectively (Fig. IX.3c).

Cruise 73-012 (25 April - 10 May 1973)

This was one of the most thorough studies of the Gulf, with detailed coverage of all zones except the upper Estuary and the region seaward of the Strait of Belle Isle. Concentrations ranged from 0.3 to 7.5 μg·L^{-1} and, when sorted geographically, the data for all zones passed the χ^2-test after transformation (there were too few data in zones E and F for meaningful tests — see Table IX.4). Combining the data for those areas where the differences between the geometric means were not significant produced a 5-compartment model (Fig. IX.3d) ; namely, the lower Estuary and Saguenay, the Gaspe and Magdalen Shallows, the northeastern region, central region and

147

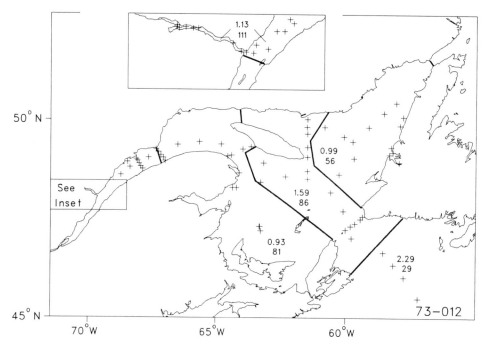

FIG. IX.3d. (*Continued.*)

TABLE IX.4. Background levels of dissolved/dispersed petroleum residues in the Gulf of
St. Lawrence. G.M. is the geometric mean concentration, and Distribution
range gives the concentrations corresponding to $\pm$ 1σ of the logtransformed
data (all in $\mu g \cdot L^{-1}$). April/May 1973 (Cruise 73-012).

Region	G.M.	Distribution range	n
A. Cabot Strait approaches	2.29	1.02 - 5.10	29
B. Cabot Strait	1.57	0.89 - 2.77	34
C. Central Gulf	1.55	0.98 - 2.46	37
D. Northeast Gulf	0.99	0.62 - 1.57	56
E. Magdalen Shallows	0.99	0.58 - 1.71	19
F. Jacques Cartier Passage	1.75	0.95 - 3.21	15
G. Gaspe Passage	0.91	0.57 - 1.44	62
H. Lower Estuary	1.85	1.10 - 3.15	78
J. Saguenay Fjord	1.51	0.86 - 2.65	33
Overall data set	1.43	0.71 - 2.85	363

the approaches to Cabot Strait. As before, the highest concentrations were seaward
of Cabot Strait and the lowest in the northeastern and southwestern regions.

As found earlier, the concentration level at the site of the *Irving Whale* did not exceed
that for the remainder of the Magdalen Shallows. Consequently, there was no evidence that leakage from the sunken barge had increased or was giving rise to elevated
concentrations of dissolved/dispersed petroleum residues in the water column. Further, concentrations in the Corner Brook/Bay of Islands area, which receives domestic
and industrial waste from the town and a pulp/paper mill, were no higher than elsewhere in the northeastern region of the Gulf indicating that these inputs were having
no noticeable effects on the general background levels in the Gulf.

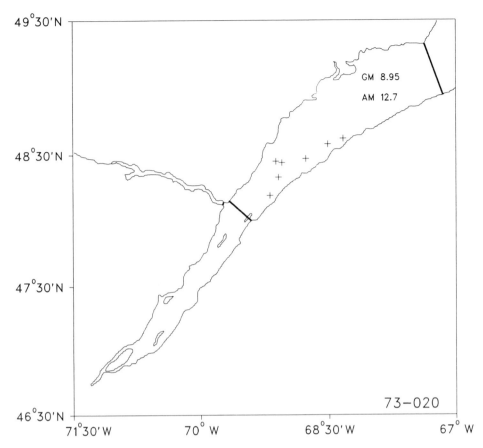

FIG. IX.3e. (*Continued.*)

Cruise 73-020 (8 July 1973)

This was a very brief survey to study the impact on the background level of dis-solved/dispersed petroleum residues in the lower Estuary from a spill of Bunker C fuel oil following a collision on July 4, 1973, between the bulk carrier, *Florence*, and the tanker, *St. Spyridon*, in the lower St. Lawrence Estuary. At the time of sampling, floating oil and associated slicks on the sea surface were being moved back and forth by the strong tidal currents in this area, but there was a net transport of oil downstream. Concentrations in the upper 50 m of the water column near the site where the incident occurred ranged from 6.5 to 95.5 $\mu g \cdot L^{-1}$, but they decreased to 5.0 to 7.0 $\mu g \cdot L^{-1}$ some 60 km downstream. Although there were too few data for a rigorous statistical treatment, the impact of the spill on the background level on the Estuary was evident. Not only were the average concentrations (arithmetic mean — A.M., 12.7 $\mu g \cdot L^{-1}$; geometric mean — G.M., 9.0 $\mu g \cdot L^{-1}$ — Fig. IX.3e) higher than those observed before the spill, but also the variability was much greater because of the inhomogene-ous distribution of the oil in the water column. Since this was a very small spill, its effects were local and short-lived and it is doubtful if it had any impact on background levels in the Gulf as a whole.

Cruise 74-028 (29 July - 15 August 1974)

Sampling during this cruise covered all areas of the Gulf except the upper Estuary and the Saguenay Fjord. Special attention was given to the approaches to Cabot Strait and the Strait of Belle Isle to compare background levels in the Gulf with those outside.

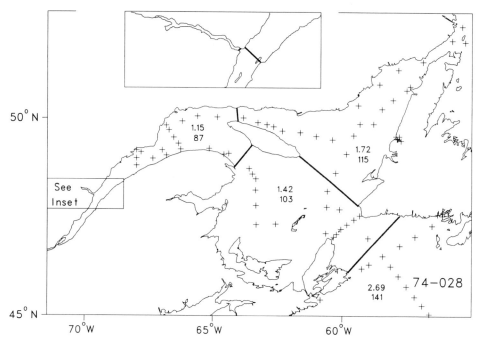

Fig. IX.3f. (Continued.)

Table IX.5. Background levels of dissolved/dispersed petroleum residues in the Gulf of
St. Lawrence. G.M. is the geometric mean concentration, and Distribution
range gives the concentrations corresponding to $\pm$ 1σ of the logtransformed
data (all in $\mu g \cdot L^{-1}$). July/August, 1974 (Cruise 74-028).

Region	G.M.	Distribution range	n
A. Cabot Strait approaches	2.69	1.44 - 5.00	141
B. Cabot Strait	1.49	0.72 - 3.07	44
C. Central Gulf	1.31	0.85 - 2.03	25
D. Northeast Gulf	1.81	0.84 - 3.91	83
E. Magdalen Shallows	1.40	0.81 - 2.43	34
F. Jacques Cartier Passage	1.51	0.63 - 3.60	32
G. Gaspe Passage	1.13	0.68 - 1.89	64
H. Lower Estuary	1.21	0.56 - 2.62	23
K. Belle Isle approaches	1.29	0.71 - 2.32	31
Overall data set	1.72	0.83 - 3.55	477

Concentrations ranged from 0.2 to 7.9 $\mu g \cdot L^{-1}$ in the Gulf, 0.5 to 13.5 $\mu g \cdot L^{-1}$ in
the approaches to Cabot Strait, and 0.5 to 4 $\mu g \cdot L^{-1}$ in the approaches to the Strait
of Belle Isle. The data sets for the entire region and the various sub-areas were lognormally distributed, and the geometric means for the various areas ranged from 1.1
$\mu g \cdot L^{-1}$ in the Gaspe Passage to 2.7 $\mu g \cdot L^{-1}$ in the approaches to Cabot Strait (Table
IX.5). In this case, a 5-compartment model consisting of the approaches to Cabot
Strait, the southern Gulf, the northern Gulf, the St. Lawrence Estuary, Gaspe area,
and the approaches to the Strait of Belle Isle (Fig.IX.3f) was appropriate.
Concentration levels were highest in the approaches to Cabot Strait, intermediate in
the central portion of the Gulf, and lowest in the St. Lawrence Estuary/Gaspe Passage
area. As in the previous cruises, the background level in the Atlantic outside Cabot
Strait was significantly higher than that in the Gulf thereby providing still further support
for the hypothesis that the Atlantic waters were the main source of dissolved/dispersed
petroleum residues in the Gulf. On the other hand, the background level in the Atlantic

outside the Strait of Belle Isle was slightly lower than that in the northeastern portion of the Gulf. The difference in the background level in the Atlantic in the Strait of Belle Isle region and that south of Newfoundland was primarily a consequence of oceanographic factors. In the former region, the water is predominantly arctic water that flows southward along the Labrador Shelf whereas that in the latter is a complex mixture of Labrador current water, slope water, and outflow from the Gulf of St. Lawrence. Thus, the similarity between the background levels in the northeastern portion of the Gulf and that in the Atlantic near the Strait of Belle Isle was a result of the exchange and mixing of Gulf and Atlantic waters in this region. In the Atlantic south of Newfoundland, however, there is much more shipping activity, and the slope water which is present in this area is thought to be a major source of the petroleum residues found here and on the Grand Banks (Levy 1983).

Cruise 75-015 (27 May - 8 June 1975)

Although sampling during this survey was focussed on the Gaspe region, coverage was also good in Cabot Strait and the lower Estuary. In addition, samples were collected from the Bay of Islands, Newfoundland, to provide further information about this area which receives domestic and industrial wastes from Corner Brook.

Concentrations were less than 1 $\mu g \cdot L^{-1}$ in nearly all samples collected during this cruise, although one sample contained more than 11 $\mu g \cdot L^{-1}$. The data from this cruise were unique in that the frequency distribution histograms for the overall data set and for the Gaspe region contained an exceptionally large number of values in the central portion of the range. Consequently, neither the untransformed nor the log-transformed data passed the χ^2-test, and it is not clear whether the arithmetic or the geometric mean is the more appropriate measure of the background level. However, in practice this is of no importance as the two means are almost identical. The distributions for the other sectors of the Gulf were lognormal wherever sufficient data were available for a meaningful χ^2-test. Not only were the background levels for the various sectors of the Gulf during this cruise remarkably uniform but they were much lower than those observed during the previous cruises (Table IX.6). As a result, a 2-compartment model (Fig. IX.3g) having background levels of 0.44 and 0.29 $\mu g \cdot L^{-1}$ emerged. Although the background levels for these two areas were statistically different, such a small difference cannot be expected to have any environmental meaning since the backgrounds themselves are well below the concentrations of dissolved/dispersed petroleum residues that are known to have immediate toxic effects or have been found to have sublethal effects on marine life (Kiceniuk and Khan 1983).

The concentration level in the Bay of Islands area of Newfoundland did not differ from that of the northeastern portion of the Gulf, and therefore, the data confirm earlier results which indicated that neither the town of Corner Brook nor its paper mill was having a detectable influence on the concentrations of dissolved/dispersed petroleum residues in the Gulf.

TABLE IX.6. Background levels of dissolved/dispersed petroleum residues in the Gulf of St. Lawrence. G.M. is the geometric mean concentration, and Distribution range gives the concentrations corresponding to ± 1σ of the logtransformed data (all in $\mu g \cdot L^{-1}$). May/June 1975 (Cruise 75-015).

Region	G.M.	Distribution range	n
B. Cabot Strait	0.40	0.23 - 0.70	62
C. Central Gulf	0.25	0.16 - 0.43	14
D. Northeast Gulf	0.29	0.18 - 0.44	26
E. Magdalen Shallows	0.41	0.29 - 0.59	24
G. Gaspe Passage	0.47	0.26 - 0.85	144
H. Lower Estuary	0.43	0.27 - 0.69	49
Overall data set	0.41		319

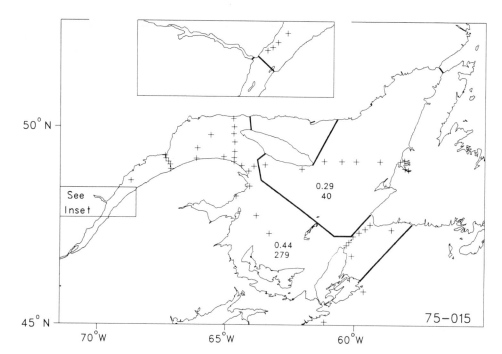

FIG. IX.3g. (*Continued.*)

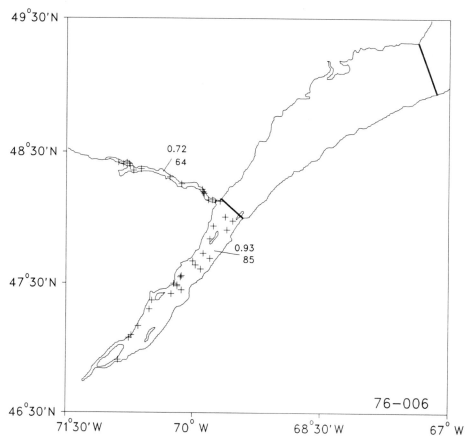

FIG. IX.3h. (*Continued.*)

Cruise 76-006 (23 - 28 April 1976)

This cruise was confined to the upper Estuary and the Saguenay Fjord and provided an opportunity to study the concentrations of dissolved/dispersed petroleum residues in the regions where fresh water from the St. Lawrence and Saguenay rivers mixes with seawater from the Laurentian Channel.

Concentrations in the Saguenay Fjord ranged from 0.35 to 8.75 $\mu g \cdot L^{-1}$ (A.M. = 0.94 $\mu g \cdot L^{-1}$; G.M. = 0.72 $\mu g \cdot L^{-1}$) (Fig. IX.3h) with most of the higher values clustered near the upper reaches of the fjord. The substantially higher concentrations (A.M. = 1.37 $\mu g \cdot L^{-1}$; G.M. = 0.92 $\mu g \cdot L^{-1}$) in this area, compared with those in the remainder of the fjord, clearly indicated the impact of local inputs of non-polar aromatic substances. Similarly, concentrations in the freshwater dominated portion of the upper Estuary were substantially higher than those farther downstream (A.M. = 2.02 $\mu g \cdot L^{-1}$; G.M. = 1.67 $\mu g \cdot L^{-1}$ versus A.M. = 0.77 $\mu g \cdot L^{-1}$; G.M. = 0.65 $\mu g \cdot L^{-1}$). Thus, in both areas, the influences of inputs associated with the river discharges were evident. These were, however, very localized because of "removal" processes (for example, the scavenging of organic substances by suspended particulate matter and its subsequent settling from the water column) associated with the mixing of fresh and saline waters. Consequently, these inputs appear to have no effect on the open Gulf.

Cruise 79-024 (25 Aug. - 2 Sept. 1979)

The sampling program on this cruise was focussed on the Estuary and Gaspe regions and, as a result, sampling was sparse in Cabot Strait and only one station was occupied in the central region.

Concentrations of dissolved/dispersed petroleum residues in the water column at this time ranged from 0.1 to 7.5 $\mu g \cdot L^{-1}$ with most of the values being less than 0.5 $\mu g \cdot L^{-1}$. Because of this clustering, the overall data set failed to pass the χ^2-test. When sorted geographically, the data for the Gaspe, upper Estuary, and lower Estuary were lognormally distributed while there were too few data from Cabot Strait and the central Gulf for meaningful tests. On this basis, the background levels were slightly greater than 1 $\mu g \cdot L^{-1}$ in the upper Estuary and somewhat less in the other regions (Table IX.7). Combining regions where the differences in background levels were not significant produced a 3-compartment model which indicated that the background levels in the upper Estuary, central Gulf, and in Cabot Strait were 1.1, 0.35, and 0.62 $\mu g \cdot L^{-1}$ respectively (Fig. IX.3i).

In addition to the water column, concentrations of dissolved/dispersed petroleum residues in the sea surface microlayer were measured at 37 locations. Concentrations ranged from 10 to over 160 $\mu g \cdot L^{-1}$, but only a few values exceeded 50 $\mu g \cdot L^{-1}$. The data were lognormally distributed with a geometric mean of 22 $\mu g \cdot L^{-1}$. That is, the general level of contamination in the surface microlayer was about an order of magnitude higher than that in the underlying water column. This is consistent with similar

TABLE IX.7. Background levels of dissolved/dispersed petroleum residues in the Gulf of St. Lawrence. G.M. is the geometric mean concentration, and Distribution range gives the concentrations corresponding to ± 1σ of the logtransformed data (all in $\mu g \cdot L^{-1}$). August/September 1979 (Cruise 79-024).

Region	G.M.	Distribution range	n
B. Cabot Strait	0.62	0.40 - 0.95	29
C. Central Gulf	0.31	0.21 - 0.47	12
G. Gaspe Passage	0.37	0.21 - 0.66	58
H. Lower Estuary	0.35	0.23 - 0.54	36
I. Upper Estuary	1.10	0.59 - 2.06	93
Overall data set	0.62	0.29 - 1.35	228

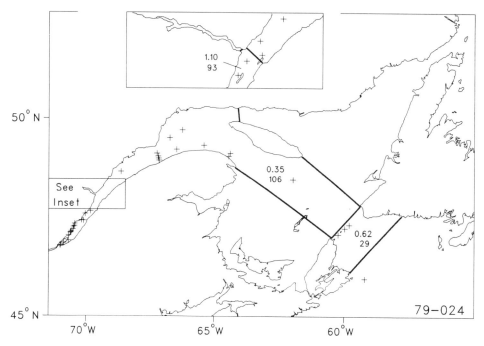

FIG. IX.3i. (Continued.)

sets of data for Baffin Bay (Levy 1979b, 1981) and the Grand Banks (Levy 1983). It is noteworthy, also, that the highest levels of surface contamination were in the Cabot Strait area and at the seaward end of the St. Lawrence Estuary.

A Comprehensive Model

The statistical analysis of data collected on 10 cruises conducted between 1970 and 1979 demonstrated that the incidence and distribution of dissolved/dispersed petroleum residues in the Gulf of St. Lawrence and adjoining marine areas during this period could be interpreted in terms of the general background levels in "boxes" of water defined on the basis of the oceanographic regimes. Since no relationship could be found between the background levels of dissolved/dispersed petroleum residues in the water column and depth or any other of the commonly measured oceanographic parameters and because the apparent differences between the concentration levels in neighbouring boxes often were not significant, the background levels in the Gulf region at the time of each cruise could be accounted for by a simple geographical model with 4 or 5 compartments. Since the general features of these models were very similar, only minor modifications were needed to produce a 4-compartment model; that is, the approaches to Cabot Strait, Cabot Strait, the Gulf, and the lower Estuary, which adequately described the background levels throughout the region for the entire period of study (Levy 1985). Water in the Gulf exchanges with both the lower Estuary and Cabot Strait, while that in Cabot Strait interacts with the Gulf and the Atlantic. The Atlantic beyond the Strait of Belle Isle was not included as no significant difference was noted between the background levels in the northeastern portion of the Gulf and the Atlantic approaches to the Strait of Belle Isle. Similarly, the upper Estuary and Saguenay regions were not included because any input to the Gulf from these regions must occur by way of the lower Estuary.

On the basis of this unified model, the general relationships between the background levels of dissolved/dispersed petroleum residues in the various sectors stood out more clearly (Table IX.8). For example, in three of the four cases where there were data,

154

TABLE IX.8. Background levels of dissolved/dispersed petroleum residues in the Gulf of St. Lawrence 1971 - 1979: Simplified model illustrating temporal changes ($\mu g \cdot L^{-1}$). Numbers in parentheses are the numbers of samples included in the geometric means.

Date	Cabot Strait approaches	Cabot Strait	Gulf	Lower Estuary
July/Aug. 1971	6.3 (3)	6.0 (21)	1.6 (200)	0.95 (18)
May/June 1972	3.0 (38)	4.6 (37)	2.9 (139)	
Feb./Mar. 1973		1.4 (33)	0.95 (78)	1.3 (7)
Apr./May 1973	2.3 (29)	1.6 (34)	1.1 (189)	1.9 (78)
July/Aug. 1974	2.7 (141)	1.5 (44)	1.45 (238)	1.2 (23)
May/June 1975		0.40 (62)	0.42 (250)	0.43 (49)
Aug./Sept. 1979		0.62 (29)	0.35 (70)	0.35 (70)

the background level in the approaches to Cabot Strait was greater than that in Cabot Strait, and in six of the seven cases the level in Cabot Strait was greater than that in the Gulf. On the other hand, there was no consistent trend between the levels in the Gulf and those in the lower Estuary.

As with the specific models, this general model endorsed the concept that the major source of petroleum residues in the Gulf is associated with the influx of Atlantic water through Cabot Strait, that inputs associated with the discharge from the St. Lawrence drainage basin are negligible by comparison, and that no substantial net exchange occurs through the Strait of Belle Isle. Within the Gulf region, the background levels in the various sectors are closely related to the major circulation pattern. Consequently, the key to managing environmental quality in the Gulf of St. Lawrence region, insofar as dissolved/dispersed petroleum residues is concerned, lies in controlling the input of these substances to the North Atlantic, particularly the approaches to Cabot Strait and, at the same time, avoiding any increase in direct local inputs.

Temporal Trends

More significant than the actual values of the background levels and their geographical distribution is the decrease in background levels that occurred during the early 1970's. The data (Table IX.8) suggested that by 1975 a steady state had been reached between the rate of input and the ability of the Gulf to assimilate dissolved/dispersed petroleum residues. While the cause of this decline is not known with certainty, it was surmised that it was brought about by measures taken during the late 1960's and early 1970's to prohibit the discharge of oil into the Gulf of St. Lawrence and to limit discharges from ships on the high seas (Levy 1985). Although discharges of oily wastes continue to occur, any reduction in the background level in the Atlantic approaches to Cabot Strait would soon be manifest by changes in the background levels in the Gulf because of the short residence time for water in the Gulf. Thus, the observed decrease in the background levels of dissolved/dispersed petroleum residues in the Gulf of St. Lawrence might indeed reflect the efficacy of pollution control measures in arresting environmental deterioration and improving environmental quality in semi-enclosed seas.

While the most recent detailed survey of the Gulf was made in 1979, data collected in 1981 at one station in the Laurentian Channel south of Newfoundland indicated that the background level in the source water for the Gulf was 0.63 $\mu g \cdot L^{-1}$ (Levy 1983). This is in excellent agreement with the level measured in Cabot Strait in 1979 and implies that the background level in the Gulf itself was probably similar, or perhaps even somewhat lower, than that observed in 1979. Thus, the environmentally positive trend towards lower levels of contamination by dissolved/dispersed petroleum residues had apparently not been reversed.

Source Material

The present background level of dissolved/dispersed petroleum residues in the Gulf of St. Lawrence is similar to those of the Grand Banks (Levy 1983) and Baffin Bay (Levy 1981) and the Hudson Strait/Labrador Shelf regions (Levy 1986). Since none of these remote areas receives large direct inputs of petroleum residues from anthropogenic sources and because the concentrations of petroleum residues throughout were very uniform, it was postulated (Levy 1985) that these substances were derived primarily from a diffuse source such as fallout from the atmosphere of non-polar aromatic substances formed during the high-temperature combustion of organic materials. Because the concentrations of dissolved/dispersed petroleum residues in the Gulf during the last few surveys were also very uniform and of the same magnitude, it was proposed that a substantial portion of the present background in the Gulf had similarly entered via the atmospheric pathway (Levy 1985). Thus, point sources such as discharges from ships or other direct anthropogenic inputs, that were probably the major source in the past, became a comparatively smaller portion of the total input during the past decade and atmospheric fallout is possibly now the predominant source.

Although small spills such as those resulting from the sinking of the *Irving Whale* and the collision between the *Florence* and *St. Spyridon* occur occasionally, their effects are usually local and short-term and have little, if any, lasting impact on the regional levels. Similarly, inputs of petroleum residues from potential land-based sources around the Gulf are presently insignificant except at the extreme upper reaches of the St. Lawrence Estuary and the Saguenay Fjord. Even there, however, the effect is highly localized because of dilution, sedimentation, and other processes of removal associated with the mixing of fresh and saline waters and, therefore, these sources do not make a significant contribution to the background level in the Gulf as a whole.

Conclusion

A 10-year series of investigations into the incidence and distribution of dissolved/dispersed petroleum residues in the Gulf of St. Lawrence has demonstrated a decline in background levels to those existing in remote, presumably pristine, marine waters of northern Canada. Since the existing background levels in all regions are well below those known to have immediate and direct adverse effects on marine life, there is no reason at present for environmental alarm. Neither, however, is there room for complacency, since any increase in the rate of input could easily alter the present situation, possibly with serious deterioration of environmental quality. As a result, monitoring of the background levels of petroleum residues in the Gulf of St. Lawrence and elsewhere should be carried out from time to time in the future.

References

AITCHISON, J., AND J. A. C. BROWN. 1969. The lognormal distribution. Cambridge University Press, London. 176 p.

GARRETT, W. D. 1965. Collection of slick-forming materials from the sea surface. Limnol. Oceanogr. 10: 602-605.

KICENIUK, J. W., AND R. A. KHAN. 1983. Toxicology of chronic crude oil exposure: sublethal effects on aquatic organisms, p. 425-436. *In* J. O. Nriagu [ed.] Aquatic toxicology. John Wiley and Sons, New York, NY.

LEVY, E. M. 1971a. The presence of petroleum residues off the east coast of Nova Scotia, in the Gulf of St. Lawrence and the St. Lawrence River. Water Res. 5: 723-733.

1971b. Impact of the Whale incident on the background levels of petroleum residues in the Gulf of St. Lawrence. Unpublished Report to Ministry of Transport, Canada.

1972. Evidence for the recovery of the waters off the east coast of Nova Scotia from the effects of a major oil spill. Water Air Soil Pollut. 1: 144-148.

1977. Fluorescence spectrophotometry: principles and practice as related to the determination of dissolved/dispersed petroleum residues in sea water. Bedford Inst. Oceanogr. Rep. BI-R-77-7: 17 p.

1979a. Intercomparison of Niskin and Blumer samplers for the study of dissolved and dispersed petroleum residues in seawater. J. Fish. Res. Board Can. 36: 1513-1516.

1979b. Concentration of petroleum residues in the waters and sediments of Baffin Bay and the Eastern Canadian Arctic — 1977. Bedford Inst. Oceanogr. Rep. BI-R-79-3: 34 p.

1981. Background levels of petroleum residues in Baffin Bay and the Eastern Canadian Arctic: role of natural seepage, p. 345-362. *In* Petroleum and the Marine Environment, Proceedings of the Petromar 80 symposium. Graham and Trotman Ltd., London.

1983. Baseline levels of volatile hydrocarbons and petroleum residues in the waters and sediments of the Grand Banks. Can. J. Fish. Aquat. Sci. 40 (Suppl. 2): 23-33.

1984. Oil Pollution in the World's Oceans. Ambio 13: 226-235.

1985. Background levels of dissolved/dispersed petroleum residues in the Gulf of St. Lawrence 1970-1979. Can. J. Fish. Aquat. Sci. 42: 544-555.

1986. Background levels of petroleum residues in the waters and surficial bottom sediments of the Labrador Shelf and Hudson Strait/Foxe Basin regions. Can. J. Fish. Aquat. Sci. 43: 536-547.

LEVY, E. M., M. EHRHARDT, D. KOHNKE, E. SOBTCHENKO, T. SUZUOKI, AND A. TOKUHIRO. 1981. Global oil pollution: results of MAPMOPP, the IGOSS Pilot Project on Marine Pollution (Petroleum) Monitoring. Intergovernmental Oceanographic Commission, UNESCO, Paris, 35 p.

LEVY, E. M., AND A. WALTON. 1971. An evaluation of two neuston samplers. Bedford Inst. Oceanogr. Rep. 71-9: 10 p.

1973. Dispersed and particulate petroleum residues in the Gulf of St. Lawrence. J. Fish. Res. Board Can. 30: 261-267.

CHAPTER X

Chemical Oceanography in the Gulf: Present and Future

P.M. Strain

*Marine Chemistry Division, Physical and Chemical Sciences Branch
Department of Fisheries and Oceans, Bedford Institute of Oceanography,
P.O. Box 1006, Dartmouth, N.S. B2Y 4A2*

Introduction

This final chapter will discuss the state of development of chemical oceanography in the Gulf of St. Lawrence. The behaviours of different chemical species in the Gulf have been described in a number of different ways. This section will examine these different interpretative approaches, asking four questions: (i) what is the present description able to tell us?, (ii) what are the limitations of the present description?, (iii) what can we learn by combining the understanding of two or more chemical species? and (iv) what are the currently unanswered questions to which we most need answers? Although what follows is organized in a way similar to the structure of the chapters in the book, it is not intended to be a summary of the material in Chapters I - IX. Rather, examples illustrating the nature of our understanding of chemical oceanography in the Gulf of St. Lawrence are selected from the descriptions of the different chemical components. In many cases the discussion will be applicable to a number of the chemical species discussed in the book.

Suspended Particulate Matter

The behaviour of suspended particulate matter in the Gulf is understood on a number of different levels (see Chapter II). On the most basic level, the SPM distribution is described for different parts of the Gulf and different seasons. From the details of this distribution, some of the mechanisms that control particle behaviour may be inferred. For example, the presence of the turbidity maximum, together with some knowledge of the dynamics of the Upper Estuary, suggested that trapping of particles by the estuarine circulation may be an important process. A more sophisticated analysis of the particle size distribution and the organic/inorganic composition of the SPM in this region then provided evidence for the importance of such a trapping mechanism, as well as showing that particle flocculation plays a role. The determination of fluxes into and out of the Gulf, and the subsequent construction of a box model for the Gulf, provide further insights into the behaviour of SPM.

These different levels of understanding yield different sorts of information. Obviously, the distributional description provides us with a picture of where and when SPM concentrations are high. The mechanistic interpretation of this data indicates what processes may be geochemically important — e.g. the circulation of particles in the Upper Estuary provides a highly efficient scrubbing mechanism for the removal of chemical species that are partitioned onto particles in the transition from a fresh to a salt water environment. The modelling of the Gulf as a single box shows the relative importance of inputs and outputs to the Gulf as a whole and can imply that important losses or gains are occurring inside the system. For example, the SPM budget in Chapter II shows that much of the St. Lawrence River's input is deposited in the Gulf, that the efflux at Cabot Strait is largely composed of organic matter produced *in situ* within the Gulf, and that internal resuspension and deposition of particulate matter accounts for $\approx 40\%$ of the total Gulf sedimentation.

These conclusions are subject to a number of limitations. Examination of inputs and outputs to the Gulf as a whole cannot yield any information about processes within the Gulf, or say anything about the nature of imbalances in budgets. For example, the distribution of a particle-associated chemical species may be controlled by the sedimentation pattern of riverborne particulates, but this pattern is not discernible from the budget. Furthermore, conclusions may be critically dependent on the quality of the input information — e.g. the estimate of deposition from resuspended sediments is derived directly from a crude estimate of the overall sedimentation in the Gulf. It is also instructive to note that a quite different conclusion was reached from an earlier model (Sundby 1974) based on less SPM data and different water flux estimates — viz. that there is no net deposition within the Gulf. Although the budget described in Chapter II does include seasonal variations, biasing of the conclusions may still result from incomplete data coverage — most of the available SPM data is for the May to November portion of the year.

From a chemical perspective, it would be useful to know some details of the behaviour of the SPM in the Gulf that the present description does not include. What is the "ultimate" fate of inorganic particulate matter delivered by rivers, and of the materials whose transport is governed by it? What is the behaviour of particles in the less dynamic deeper waters of the Gulf and Estuary, and how do such processes influence the behaviour of materials which cycle between the surface sediments and the water column (e.g. Mn, Co, silicate)? Are there geochemically important transformations between particulate and colloidal reservoirs? How does internal particle cycling in the water column influence, and how is it influenced by, the biota?

Nutrients

There are a number of parallels between the existing description of nutrient behaviour in the Gulf and that of SPM. A large amount of distributional data is available for silicate, phosphate and nitrate; nutrient distributions have been used to discern processes important to nutrient geochemistry; a budget has been calculated for the whole Gulf system (see Chapter III). An additional approach that has been used to elucidate nutrient processes is the correlation of nutrient concentrations with other physical/chemical parameters. In the Upper Estuary, careful determinations of nutrient concentration–salinity relationships have revealed important changes in the very low salinity regime, and stress the importance of the proper identification of endmembers in estuarine mixing studies. Relating nutrient concentrations in the deep water of the Laurentian Channel to the concentrations of dissolved oxygen and the carbon isotope composition of the dissolved inorganic carbon explains some features of the regeneration of inorganic nutrients in Gulf deep waters.

The nutrient budget for the Gulf is significantly out of balance for both nitrate and silicate. There are obvious rationalizations for these imbalances — viz. that silicate is lost to the Gulf sediments and that non-nitrate forms of nitrogen play important roles in the Gulf's nitrogen cycle. However, these conclusions are subject to the quality of the model inputs and to the assumptions used in its construction. Seasonal coverage of nutrient measurements is as good as for any chemical parameter, but there still is a shortage of winter data. It was necessary to pool data from many years, with different seasons sampled in those years, to produce the budget. Obviously such models cannot consider interannual variability or processes occurring on time scales more rapid than seasonal.

A second major limitation of the budget is the assumption that fluxes through the Strait of Belle Isle are small. Recent physical oceanographic studies of the area show that water fluxes through the Strait are not negligible (Chapter I). These considerations may be particularly important for nutrients. Not only do significant transports through the Strait of Belle Isle undermine the assumptions El-Sabh used in his calculations of flows through Cabot Strait, but the possibility that Labrador Shelf water is an important component of the intermediate layer in the Gulf may mean that an important input of nutrients to the Gulf has been neglected. There is probably insufficient data available to determine whether important concentration gradients exist in the northeastern sector of the Gulf. This region of the Gulf has often not been included in field programs because of its relative remoteness, its long ice season, and the apparent lack of relevance to the highly productive more southern sector of the Gulf. In the design of any new sampling programs, the details of the region's dynamics, such as the seasonality of exchanges through the Strait of Belle Isle and the location of the inflow along the north shore of the Gulf, should be considered.

A logical extension of the present budget would be the construction of more sophisticated box models that subdivide the Gulf, giving them the potential to reveal details about internal processes. Bewers and Yeats (1983) have constructed such a model for nutrients, in which they subdivide the Gulf both vertically and horizontally (Fig. X.1). Additional inputs are required for such a budget, such as water fluxes at the internal boundaries and the composition and fluxes of particles between the surface and deep layers. Since phosphate was in balance in the Gulf as a whole, Bewers and Yeats used it to tune some of these input parameters within reasonable ranges. Applying the resulting model to silicate and nitrate showed that the silicate imbalance is confined to the deep layer, providing evidence that the sedimentation explanation is correct, and that the nitrate imbalance is a surface layer phenomenon. Furthermore, it was possible to place constraints on the concentrations of reduced nitrogen species that would be required to account for the differences.

The additional conclusions provided by this more sophisticated model are interesting and worthwhile, but they are nearing the limit of what can be gained by this approach using available data. Care must be taken that the data used are capable of supporting the conclusions reached. An important consideration in this evaluation is the density of sampling in both time and space. In addition, there is no point in creating more compartments than there are identifiably discrete environments.

Much of what we would like to know about nutrient behaviour in the Gulf relates to processes that occur on time scales, both short and long, that cannot be examined using existing data. For example, inputs from the St. Lawrence River not only show considerable seasonality, there are also intriguing differences in the scatter of phosphate concentrations from season to season. The scatter present at some times of year is likely due to processes with time scales too short (days to weeks) to be described by historical sampling strategies.

Although present distributional data has been used to examine the regeneration of nutrients in deep waters, this data is of very limited use in the study of the much more rapid cycling of nutrients in surface waters. Phosphate and nitrogen may be reused

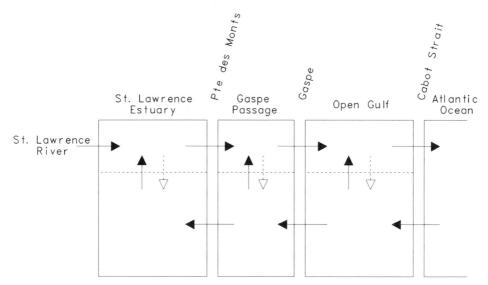

FIG. X.1. Enhanced box model for nutrients in the Gulf of St. Lawrence. Solid arrows indicate advective fluxes; open arrows indicate particulate fluxes.

many times prior to settling out of the biologically active surface layer. Understanding this recycling is important for the study of primary productivity. High density sampling for phosphate and nitrate might elucidate some aspects of this surface behaviour, but it would be difficult to distinguish advective features from biochemical activity. The study could be further complicated by a very rapid cycling of nutrient through a relatively small standing stock — high, and nearly balanced, uptake and remineralization rates are compatible with a low, relatively invariant, surface nutrient concentration. Details of the nutrient cycles might only be available from more direct rate measurements. Close examination of the interrelationships between phosphate and nitrogen may also be useful. Since nitrogen is thought to be growth limiting in the Gulf, it is particularly important to understand its behaviour. Nitrogen cycling undoubtedly involves reduced nitrogen forms such as ammonia, urea and other organic compounds about which almost nothing is known in the Gulf. Field work to examine these constituents of the nitrogen cycle should be a high priority. Studies of both phosphate and nitrogen must be designed in view of the short time scales important in the cycling of these nutrients.

The existing data base also contains little information about longer term variations that may occur in nutrient dynamics in the Gulf. Some consistency has been observed in nutrient gradients in the deep water of the Laurentian Channel over the period 1971–75. However, the interannual variability of important aspects of nutrient behaviour, such as inputs through Cabot Strait and the Strait of Belle Isle and the interaction between freshwater input and the supply of nutrients to the surface layer, requires further study.

Organic Carbon

The organic carbon data base for the Gulf of St. Lawrence is as complete as that for any of the other chemical parameters. Particulate and dissolved organic carbon and particulate organic nitrogen have been sampled in detail at many times of the year in Cabot Strait, and have been monitored in St. Lawrence River discharge over a five year period. This extensive data base made it possible to take a more rigorous approach to the construction of the budget described in Chapter IV than has been done with other chemical species. Organic matter concentrations in Cabot Strait were mapped onto water flux estimates at nearly the same resolution in both time (almost monthly) and space (25 m slices down to 250 m depth) that El-Sabh (1977) used in the water flux calculations. The extensive data base also makes it possible to do a sensitivity analysis of some of the assumptions implicit in the budgets for SPM, nutrients and organic carbon.

The 3-month period of June, July, and August has been chosen to investigate the budget because both water flow and organic matter data are available for all 3 months. The budget calculates the chemical mass flux as the product of the water flow and organic concentration, assuming that both are constant for the month. In Table X.1 the results of this calculation of the July flux profile are compared with one derived by integrating the instantaneous water flow-concentration product (the instantaneous values were assumed to lie along straight line segments joining the values for mid-June to mid-July to mid-August). The differences between the two calculations are almost all below 10 %, with larger errors confined to very small fluxes. Another inherent assumption is that the mean values for water flow and organic matter concentrations represent the conditions at the same time. Since El-Sabh's calculations are based on several years of data, his average values have been assigned to the middle of each month. In contrast, the organic matter data for a given month might be based on sampling from a single cruise at either end of the month. In Table X.2 the original July mass fluxes are compared with those calculated by assuming that the concentration varied linearly with time in July and either (i) July and August concentrations apply to the first day of each month or (ii) June and July concentrations apply to the last day of the month. The differences between these calculations and the budget calcula-

TABLE X.1. Particulate organic carbon fluxes (kg·s^{-1}) through Cabot Strait in July. The fluxes used in the model in Chapter IV ($\overline{m \cdot c}$) are compared to fluxes calculated by integrating water flows and concentrations assumed to vary linearly with time over the period from June to August (linear).

Depth	Out			In		
(m)	$\overline{m \cdot c}$	linear	% diff	$\overline{m \cdot c}$	linear	% diff
0–25	14.29	15.05	5.3	4.06	—	—
25–50	6.35	6.44	1.4	3.79	3.42	9.8
50–75	1.94	2.03	4.6	3.03	2.70	10.9
75–100	1.06	—	—	1.87	1.71	8.6
100–125	0.75	—	—	1.28	1.17	8.6
125–150	0.47	—	—	0.77	0.73	5.2
150–175	0.27	0.25	7.4	0.48	0.53	10.4
175–200	0.11	0.11	0	0.33	0.38	15.2
200–225	0.05	0.06	20	0.25	0.32	28
225–250	0.02	—	—	0.17	0.25	47
250–450	0.002	—	—	0.63	—	—

TABLE X.2. Particulate organic carbon fluxes (kg·s^{-1}) through Cabot Strait in July. The fluxes used in the model in Chapter IV ($\overline{m \cdot c}$) are compared to fluxes calculated by offsetting the concentrations relative to the water flows by ≈ 0.5 month. Water flows are assumed to be constant through the month; concentrations are assumed to vary linearly with time through July and to be valid for either the first or last day of the month.

Depth	Out					In				
(m)	$\overline{m \cdot c}$	First Day	% Diff	Last Day	% Diff	$\overline{m \cdot c}$	First Day	% Diff	Last Day	% Diff
0–25	14.29	13.93	2	14.77	3	4.06	4.13	2	—	—
25–50	6.35	5.11	20	7.21	14	3.79	3.70	2	3.88	2
50–75	1.94	1.57	19	2.51	29	3.03	2.49	18	3.35	11
75–100	1.06	0.89	16	—	—	1.87	1.44	23	2.27	21
100–125	0.75	0.68	9	—	—	1.28	1.03	20	1.42	11
125–150	0.47	0.45	4	—	—	0.77	0.64	17	0.82	6
150–175	0.27	0.23	15	0.35	30	0.48	0.42	13	0.52	8
175–200	0.11	0.09	18	0.14	27	0.33	0.29	12	0.37	12
200–225	0.05	0.04	20	0.07	40	0.25	0.24	4	0.32	28
225–250	0.02	—	—	0.02	0	0.17	0.16	6	0.24	41
250–450	0.002	—	—	0.002	0	0.63	0.54	14	0.71	13

tions are larger than in the previous test, reaching values up to 20 % even for the important fluxes. However, these potential errors are no greater than the precision expected in this kind of budget calculation. Similar results are obtained using the particulate organic nitrogen data.

The somewhat surprising robustness of the organic carbon budget is due in part to the nature of the organic matter distribution. Concentrations of organic carbon are highest, and most stable, in the highest water flows, which are also relatively stable through this three month period. The mass fluxes in these surface flows dominate the total exchange through Cabot Strait. Budgets will be equally well constrained for any component when its flux is not changing rapidly with time. This will likely be the case for materials with relatively invariant surface concentrations which are mostly associated with the freshwater outflow at Cabot Strait, such as SPM. Where deep and intermediate water processes are important, however, such calculations will probably be less stable due to the higher variability found in the flows at depth. Such is the case for materials like the nutrients, whose inputs at Cabot Strait exceed exports.

TABLE X.3. Water flows (10^3 m^3 s^{-1}) through Cabot Strait in June. Flows calculated from data collected over the period 1957-66 are compared to flows calculated using 1962 data only. Data from El-Sabh (1977). Surface outflows in the top 50 m have been adjusted using data in Bugden (1981).

Depth	Out			In		
(m)	1957-1966	1962	% diff	1957-1966	1962	% diff
0-25	199.4	231.1	16	0	24.5	—
25-50	75.6	158.3	109	12.7	38.3	202
50-75	10.1	61.7	511	18.7	43.8	134
75-100	0	24.5	—	22.1	44.5	101
100-125	0	9.0	—	24.3	59.9	147
125-150	0	2.0	—	21.5	31.3	46
150-175	2.9	0	100	17.3	24.0	39
175-200	4.8	0	100	13.2	19.4	47
200-225	6.7	0	100	11.4	15.2	33
225-250	6.3	0	100	8.9	12.9	45
250-450	9.4	0	100	49.2	54.6	11

El-Sabh (1975, 1977) presented geostrophic water flow calculations for the month of June derived in two different ways. One set was based on average hydrographic properties over the period 1957-66; the other was based on six hydrographic sections occupied in June 1962. These different water flows are compared in Table X.3. The 1962 data yield flows as much as five times greater than the longer term average; the flows in the 25-75 m deep outflow are very much greater than those based on the 1957-66 data. Using the 1962 data results in a 60 % increase in gross carbon export and a 31.5 % increase in net export.

To date, the chemist's use of El-Sabh's water flow estimates has been rather uncritical, but there has been no other physical information that was so amenable to the calculation of chemical fluxes. Consideration of the surface outflow through Cabot Strait illustrates just how fragile these calculations may be. The exports for materials with highest concentrations in surface water will be dominated by the water flow in the top 50 m, which is more than 1.5 times that in the rest of the water column. As noted in Chapter I, Bugden (1981) considered El-Sabh's estimates too low because of the lack of data near the south side of the Strait, and adjustments have been made in these flows prior to their use in the chemical budgets in an attempt to correct for this deficiency. This adjustment amounts to ≈ 35 % of the flow in June. An evaluation of the accuracy of El-Sabh's calculations is necessary to determine the accuracy of present chemical budgets for the Gulf. In addition to the limitations of the geostrophy, this evaluation must also consider the sensitivity of these calculations to the significant and variable water flows and the potentially variable salt supply through the Strait of Belle Isle (Petrie et al. 1988). Improvements in budget and flux calculations will require evaluations both of the variabilities of the concentrations of chemical species and of the water flows most responsible for the fluxes of those species. Such evaluations will make it possible to construct budgets in which the physical and chemical data are coherent and to define over what time scales such budgets are valid.

Stable Isotopes

The measurements of stable isotope ratios described in Chapter V differ somewhat from the other measurements described in this book in that there is essentially no intrinsic interest in the ratios themselves. Unlike an SPM concentration, which might indicate the likelihood of a metal partitioning from the dissolved to the particulate phase, or a nutrient concentration, which might reveal whether conditions are right for a plankton bloom, the value of stable carbon or oxygen isotope measurements lies in what their variations can reveal about sources or transformations of material. Stable isotope and other indirect measurements capable of yielding insights on the chemical environment will be increasingly important to chemical oceanography studies because they provide a different perspective on chemical mechanisms.

Since stable isotope measurements are process oriented, the determination of their distributions is a first step in the evaluation of their potential usefulness. Organic carbon isotope ratio distributions have led to insights into such things as the sources of sedimentary organic carbon and the interaction of suspended particulate matter with both sedimentary organic carbon and planktonic organic matter. Since these measurements have been made on bulk organic carbon, the understanding that has resulted is general in nature. Future studies of the geochemistry of both natural and anthropogenic organic materials must include the examination of the behaviour of individual classes or individual organic compounds. Selecting the important compound types from the bewildering suite of organic compounds will not be simple, but potentially interesting compounds will include both those of interest in themselves and ones which may reveal details of the chemical environment.

As pointed out in Chapter V, stable oxygen isotope measurements might be able to contribute to an understanding of the importance of northern waters to the water masses of the Gulf, although such a study would push the current analytical technique to the limit of its precision. Such measurements might help to assess the distribution of waters entering the Gulf through the Strait of Belle Isle.

Trace Metals

The current picture of trace metal behaviour in the water column is derived largely from the evaluation of the interrelationships between dissolved and particulate metal concentrations and indicators of the processes which control their distributions. Metal–salinity relationships, which indicate the degree to which simple mixing of fresh and salt water determine concentrations, have been the starting point for understanding most of the metals in the water column. For a metal or a metal phase showing non-conservative behaviour, further information on the processes controlling its distribution has been found in metal–nutrient relationships (for biologically cycled metals), in dissolved/particulate covariances (for metal cycles with important redox or adsorption components) or in comparisons of sediment and water column observations. The overall result of these analyses is a picture of what processes are important in the control of trace metals which are removed from the water column in transit through the Gulf, in what regions of the system these processes are most active, and to some extent what the seasonal variabilities of these processes are.

There are at least two questions that follow logically from the focus of these studies on the fate of trace metals in the coastal zone: (i) what implications do these processes have on the supply of materials to the deep ocean? and (ii) what implications do these studies have on coastal zone management requirements?

Bewers and Yeats realized that the Gulf system, with its restricted connections with the Atlantic and the large distances between its principal inlets and outlets, could serve as an excellent setting in which to measure the transports of materials through the coastal zone into the open ocean. They combined El-Sabh's (1975) geostrophic calculations with their water column trace metal measurements on the Gulf to construct a global model predicting the residence times of trace metals in the world's oceans (Bewers and Yeats 1977). These calculations, which compared the total oceanic reservoir of a metal with its rate of input, are significantly altered by the proper correction of river flux data for the removal that occurs in the nearshore. In doing these calculations, Bewers and Yeats distinctly improved the agreement between the oceanic residence times predicted by this calculation and ones based on deep ocean sedimentation rates. This improved agreement is evidence both that the coastal processes have been described with reasonable accuracy and that the Gulf is a useful model for the world's coastal zones. The results from this model and a subsequent one also based on Gulf of St. Lawrence data (Yeats and Bewers 1983) showed that the oceanic inputs of Zn and Cd are considerably in excess of their removal through sedimentation. Although the Zn excess could have been due to the quality of the data then available, the Cd excess appeared to be real. Other calculations supported the assertion that the mobilization of Cd due to man's activities is now comparable to that occurring naturally.

Yeats et al. (1978) analyzed the sensitivity of the Gulf of St. Lawrence to increased anthropogenic discharges of trace metals (i.e. those due to man's activity). They found that the signal resulting from riverborne inputs of trace metals is reduced significantly by removal processes occurring in estuarine waters. In addition, variability in measured coastal zone concentrations resulting from natural variability and analytical imprecision significantly limits one's ability to detect changes in these concentrations. As a result, they concluded that river monitoring would produce more reliable indications of changing environmental quality. This approach should be valid for most other riverborne contaminants as well, but it may not be valid for coastal areas which receive inputs from a number of different significant sources.

In addition to determining sensible monitoring strategies and providing information on which metals have undergone a general increase in mobilization due to man's activities, experience in the Gulf of St. Lawrence has also shown that the geochemical expertise from marine chemical studies is essential in dealing with acute pollution problems. The lobster fishery in Belledune Harbour on Chaleur Bay was closed in 1980 after high levels of Cd were found in lobster tissues. A geochemical survey found that there were elevated levels of Cd in waters and sediments of the harbour and adjacent areas that originated from a nearby lead smelter. Remedial waste treatment at the smelter led to a significant reduction in the pollution of the area and the fishery was subsequently reopened (Uthe et al. 1986). The rapid identification of the source of the problem, and the initiation of corrective action, were critically dependent on an appreciation of the regional levels of Cd in both the water (Chapter VI) and the sediments (Chapter VII) in the absence of local sources and of the mechanisms responsible for the distribution of Cd in coastal waters.

Mercury pollution in the Saguenay Fjord provides another example of how basic geochemistry is important in the resolution of environmental concerns. As described in Chapter VIII, Loring (1975) discovered abnormally high levels of mercury in the sediments of the Saguenay Fjord. An examination of the distribution of Hg in the region indicated that a chlor-alkali plant on the Saguenay River was the source of most of this contamination. Loring and Bewers constructed a mercury budget which predicted that recovery of the water in the Saguenay system would be relatively rapid, perhaps as short as two years, whereas recovery of Saguenay sediments would require a much longer period. The balancing of this budget required a surprisingly large flux of mercury through the Saguenay in 1973, 2 years after discharge regulations for chlor-alkali plants were imposed. Loring and Bewers were justifiably cautious in not explicitly explaining this flux. Non-compliance with the discharge limits or remobilization of locally-retained mercury were potential explanations. Simultaneous determination of the sedimentation rate and the mercury concentration in sediment cores subsequently showed that the mercury contamination began at the same time as the start-up of the chlor-alkali plant, and that mercury fluxes into the sediments fell off sharply near the 1971 horizon in the cores, consistent with industry complying with government regulations. The high mercury fluxes in 1973 were due to the remobilization of mercury from sediments in the Saguenay River that had been contaminated prior to 1971.

The determination of sedimentation rates referred to above was itself made possible by an understanding of the geochemistry of ^{210}Pb. It was first realized in the early 1970's that ^{210}Pb could be used to date sediment cores from some coastal environments. It was further realized that the sedimentary regime at the head of the Saguenay Fjord was an ideal site to apply this technique. The accurate geochronologies that have been determined from Saguenay cores made it possible to determine the history of a number of processes in the Saguenay Fjord and its drainage basin. In addition to that of mercury, the histories of pollution due to pulp mill waste and polycyclic aromatic hydrocarbons have been determined. Time lags in the appearance of bomb-derived radionuclides in Saguenay sediments have made it possible to model the transport of metals through the Saguenay drainage basin to the Fjord. The resulting drainage basin model describes the interaction of the land and the Saguenay Fjord in a way analogous to the description of the Gulf and the Atlantic Ocean inherent in the Gulf budgets.

Petroleum Hydrocarbons

Research into petroleum hydrocarbons at the Bedford Institute was initiated as part of the response to the oil spill from the tanker *Arrow*, which ran aground on the Atlantic coast of Nova Scotia in February, 1970. It was immediately recognized that the geochemical framework necessary to evaluate the impact of the spilled oil was not available. In the summer of the same year, the first of many cruises which included measurements of petroleum residues in the Gulf of St. Lawrence was conducted. The Gulf was surveyed at this time so that petroleum levels in a similar area not affected by the spill could be compared with measurements from the spill zone.

Research into petroleum residues has continued to be a mixture of responses to environmental emergencies and projects that investigate the geochemical behaviour of oil in the environment. The latter has resulted in a long term record (10 years) of the concentrations of petroleum residues in the Gulf (see Chapter IX). This record shows that concentrations have declined in the Gulf. It has also been possible to assess the impact of oil at two accident sites — the resting place of the sunken oil barge *Irving Whale* and the site of the collision between the *Florence* and the *St. Spyridon* — in terms of the levels observed in the long term monitoring study.

This monitoring program represents a conscious effort to directly measure the variations in a chemical's concentration over a long period. Although there is some fortuitous information on other long term trends in the Gulf, such as the consistent gradients in nutrient concentrations in deep Laurentian Channel water observed from 1971 to 1975, the St. Lawrence River discharge is the only other part of the Gulf to be intentionally monitored over a long period. Such long-term programs can provide useful information, but they require a considerable commitment of resources, they may be specific to one chemical species, they may suffer from changes in analytical technique and they may in the end not reveal anything about the mechanisms responsible for the trends observed. There are other ways to study such long term changes that may be available for some components. In some locales, measurements made on well preserved, well dated sediment cores have proven very useful in determining the history of pollution inputs over periods of several decades (Chapter VIII). Understanding the geochemical mechanisms that control distributions of chemicals in the marine environment so that trends may be predicted can minimize the need for monitoring programs for many chemical species.

Chemical Interactions

The organization of both this book and most of this chapter might give the impression that the interpretation of the behaviour of each chemical species has been done independently. Much can be learned from a consideration of how the different chemical components influence each other, of how their responses to changes in the chemical, physical or biological environment are similar or different, or of what the behaviour of one component indicates about the environment of another. An integrated approach is fundamental to many of the studies described in this book.

The interrelationships of the budgets for SPM, nutrients, and organic carbon are good examples of how integrated data interpretation provides more insights into geochemical pathways than would isolated evaluation of data for each component. An imbalance in the silicate budget has been explained as the amount of silicate that is being permanently incorporated into the sediments. This flux can then be subtracted from the SPM flux into the sediments to yield the sediment flux that must be due to other sources. Since organic matter concentrations are low in sediments, most of the deposition is of inorganic material. This integrated approach was also used in the determination of the sources and sinks of organic carbon in the Upper St. Lawrence Estuary (Chapter V). A program which simultaneously determined C/N and carbon isotope ratios was able to resolve a discrepancy between earlier isolated interpretations of the two types of data, and in doing so provided additional information on the carbon cycle. ^{210}Pb studies of the sedimentary environment in the Saguenay Fjord have not only made it possible to describe the sedimentary history of the Fjord in great detail, they have also provided a time scale for sediment cores which have subsequently been used to determine the depositional history of several important chemical constituents.

Dissolved oxygen and alkalinity are two chemical parameters which may be useful indicators of variations in the chemical environment. Both these parameters have been extensively measured in the Gulf (see Appendix). The dissolved oxygen data has received comparatively little attention, other than the relationship between oxygen and nutrient concentrations in deep Laurentian Channel water (Chapter III); no attempt has been made to integrate the alkalinity data with the description of dissolved oxygen, nutrients and the isotope ratio of the dissolved inorganic carbon.

Just as it is important for physical and chemical oceanographic information to be coherent in the formulation of budgets, it is important that the data used to examine the interrelationships between two chemical constituents be compatible. If the distribution of one component is patchy in space or variable in time, then this variability may obscure a relationship with another chemical species for which measurements were made at a different time. The optimum situation is one in which the same water sample is analyzed for both components. Figure X.2 shows the percentage of organic carbon in suspended particulate matter in three cross-sections across the Laurentian Channel, based on the analysis of the same water samples for both POC and SPM. This figure shows the following three trends: (i) the organic content of SPM is lower in deep water than in surface water, (ii) the organic content in SPM in surface water increases with distance from Pte des Monts, and (iii) the organic content in surface SPM is lower on the south sides of the Gaspe and Cabot Strait sections where the influence of the St. Lawrence River discharge is the strongest. The surface layer pattern is created by the dilution of the *in situ* produced organic matter by inorganic material supplied by the St. Lawrence and other rivers. It was noted earlier that the SPM data alone was not adequate for the determination of how riverborne material was distributed through the Gulf. The joint interpretation of SPM and POC data shows, however, that riverborne inorganic matter is present in the surface outflow at Cabot Strait and is therefore found throughout that part of the Gulf influenced by the St. Lawrence discharge.

Conclusions

A number of general themes emerge from this discussion of chemical oceanography in the Gulf of St. Lawrence. The existing picture of chemistry in the Gulf describes only the coarsest and most obvious time and space scales and frequently only the most obvious chemical components. It must be emphasized that this comment is not a criticism of the work that has been done so far, but indicates that it is only the beginning. Its limitations were in some instances dictated by the state of the art, and in others by the availability of resources and logistics. Frequent reference has been made in this chapter to the inadequacy of existing data for the study of some important chemical processes. However, there is still useful information to be gleaned from the present data base, and there are several studies underway interpreting these data.

One important limitation of existing data is that the Gulf has not been sampled sufficiently frequently in time or space to properly characterize important scales of variability. It is important to be aware of the restrictions that this density of sampling impose on the interpretation of the data collected. These same restrictions are likely to apply to any new sampling program using traditional survey techniques. A first step to an

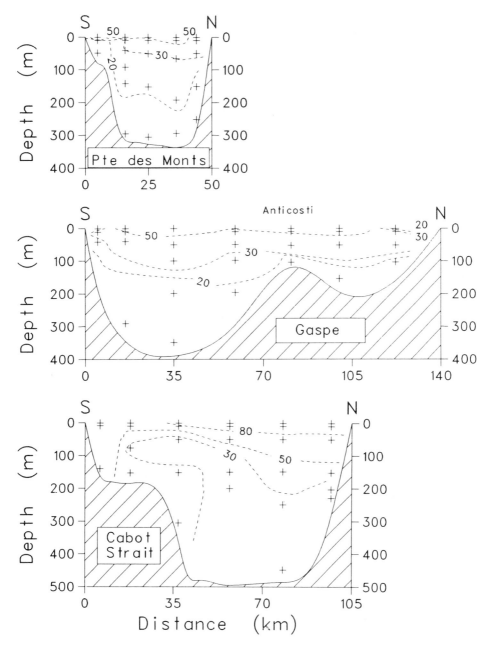

FIG. X.2. The percentage of particulate organic matter in suspended particulate matter in three cross-sections of the Gulf. (All data from cruise 75-015).

improved understanding of marine chemical processes in the Gulf will be the imaginative design of field programs.

The description of properties varying in time has been restricted to those that could be observed by sampling programs which would have typically required two weeks to survey a large area, which on average might have been conducted during two different seasons, and which might or might not have been repeated for the same season in different years. Obviously, such measurements cannot reveal much about the interactions between surface chemistry and plankton blooms, cannot evaluate the influence of the spring freshet on all parts of the Gulf, cannot distinguish between long term

169

trends and interannual variability, to choose only a few of the more obvious examples. In addition, it has frequently been necessary to interpret all available data as a homogeneous set just to have an adequate quantity of data. In a system as dynamic as the Gulf, such data treatment has inevitably produced a blurred picture.

Information gaps on distributions in space include the undersampling of areas such as the northeastern sector of the Gulf, which is assuming increased importance in the light of the recent physical oceanographic work, and the Magdalen Shelf, which is the part of the open Gulf most strongly influenced by the St. Lawrence River signal. There has also been undersampling of the surface layer, in both space and time, so that detailed descriptions of biologically cycled components and understanding of the mechanisms responsible are not possible. Processes in the surface water may be rapid, may exhibit diurnal or other short-term cycles, and may be patchily distributed in space. The study of such mechanisms presents formidable difficulties in experimental design. In theory it would be possible to address some of the processes that take place on these scales by intensive application of traditional sampling strategies. However, limited resources make approaches like multi-ship surveys or frequent resurveying impractical alternatives.

Given these limitations, future advances will depend on the design of field experiments rather than field surveys. Some of these experiments will likely involve more intensive study of a more limited area than has been the case in past field programs. The program focussing on sediment-water interactions in the Lower Estuary currently underway at the Institut Maurice Lamontagne is an example of this more sophisticated approach. The approaches to be taken could include the measurement of additional chemical species to more fully examine a geochemical pathway, or increased dependence on evaluating the covariances between different variables, or making measurements that yield direct information on the rates of processes rather than the concentrations of components. Examples of the first of these suggestions are the previously mentioned analyses of reduced nitrogen species and individual organic compounds or compound classes. It may be necessary to consider additional sources of material (e.g. comparatively little attention has been given to atmospheric inputs of chemicals to the Gulf). It may also be appropriate to examine different chemical phases. Recent geochemical thinking suggests that colloids may play important roles in the geochemistry of both trace organic and trace inorganic constituents. Some characterization of metal-organic interactions in colloidal phases is available, but both better methods and more field-based research are required to understand the importance of this fraction in areas like the St. Lawrence Estuary. Sampling this fraction is technically very difficult, but the recent applications of high volume filtration techniques to the separation of the colloidal phase in seawater show promise (Whitehouse et al. 1986).

Many chemical processes in the surface layer will be biologically mediated. Almost all of the chemical oceanography described in this book has been done in relative isolation from other marine science disciplines. Understanding surface water chemistry in detail will require more cooperation with biological oceanographers so that chemical measurements may be more directly related to biological observations than is possible using data based on separate field programs.

A number of comments through this chapter have referred to a similar lack of coherence between chemical and physical oceanographic measurements. The budgets described in this book have been constructed by chemical oceanographers using whatever physical oceanographic information was best suited to their purposes. This approach has introduced uncertainties in the compatibility of the chemical and physical data. Further refinement of budgets in the Gulf will require better liaison between chemical and physical oceanographers. It will be as important to assess the scales of variability in water fluxes, sources, and water mass characteristics as it is to assess the chemical variability. Since the creation of such improved budget/flux calculations would require considerable effort, it is important to ask whether such additional accuracy is desirable. Since such an exercise may not produce substantial new insights into the geochemical mechanisms that control chemical species, this effort may only be warranted when the chemical fluxes are important for the understanding of other processes, e.g. as nutrient fluxes are for the understanding of primary productivity.

How does this description of the current understanding of chemical oceanography in the Gulf of St. Lawrence relate to the concerns of those responsible for the protection of the marine environment in the Gulf and the management of its living and non-living resources? The scientific capabilities that such environmental managers require to meet their obligations include: (i) the ability to respond to environmental emergencies, (ii) the ability to forecast trends, (iii) the ability to assess the viability of commercial fisheries, and (iv) the ability to assess the viability and impacts of aquaculture projects. What chemical oceanographic studies are necessary to these goals, and how are the concepts discussed in this chapter related to them?

As described earlier, the ability to respond sensibly to environmental emergencies depends on the geochemical expertise necessary to put the emergency in context. At Belledune Harbour, the knowledge of natural Cd distributions and behaviour in the marine environment was needed to determine the extent of the cadmium contamination. In a similar way, the impact of oil leaking from the sunken barge *Irving Whale* could only be evaluated in the larger context of the distribution and behaviour of oil in the Gulf (Chapter IX). The study of mercury pollution in the Saguenay Fjord is another example of the application of geochemical expertise to an environmental problem. Our current level of sophistication has equipped us reasonably well to deal with these severe, localized problems when they involve certain inorganic contaminants or petroleum hydrocarbons. However, continuing geochemical studies are required to extend this capability to other inorganic materials — lead and tin are two metals with potential environmental significance on which additional information may be required. Comparatively little is known about the geochemical behaviour of individual organic compounds or even classes of compounds. The work described in Chapter IV on polychlorinated biphenyls (PCB's) and the DDT group of pesticides might best be classified as "incidence" surveys. Even such preliminary work is not available for other man-made organic contaminants (including materials like the chlorinated dioxins, furans and camphenes). A better understanding of the way in which such materials are distributed in the environment, and the processes that control their distributions, is required before the prediction of the fate of these compounds in the environment is possible. Better understanding of the behaviour of some of these materials is currently awaiting the development of improved analytical techniques. Studying the pathways of such materials in the environment requires the capability to measure ambient concentrations at levels found at relatively uncontaminated sites on relatively large numbers of samples.

The ability to forecast trends is required both to address the question of whether there is a degradation of the marine environment in the Gulf of St. Lawrence occurring on a large scale over long periods of time and to evaluate the effects of natural fluctuations on the productivity of the Gulf. As noted earlier, monitoring for changes due to widespread trends in anthropogenic releases may be best carried out on river discharges and other inputs (e.g. atmospheric) to the Gulf, where such changes are least likely to be obscured by the dynamics of the system. Such monitoring must distinguish between the natural variability of the system and a long-term change due to anthropogenic inputs. It was also pointed out that monitoring may be an inefficient or even inadequate strategy for following changes in environmental quality. Geochemical studies offer the best hope of predictive capability, but some monitoring activity will be required to verify the geochemical models.

In addition to these concerns about a possible large scale deterioration in environmental quality in the Gulf, it will be necessary to assess the impact of locally important contaminant sources. Because past studies have focussed on large scale features of chemical behaviour, it is more difficult to assess these local problems from our current understanding of geochemical processes. Studies of the chemical dynamics of small harbours, bays, estuaries and other coastal sites would provide a valuable framework for dealing with local environmental problems. Such research should include consideration of the rates and mechanisms of the transport and removal of contaminants. It will still be necessary to assess the impact of individual contaminants and the importance of local sources at specific sites, but the general information from these geochemical studies should minimize the effort required.

Natural fluctuations in physical and chemical conditions in the Gulf could have implications for the viability of commercial fisheries. From a chemical perspective, it will be necessary to investigate nutrient dynamics on time and space scales currently poorly understood — short and long term variability in time and small scale dynamics in inshore locales. Variations in nutrient supply to the Gulf will have implications for the overall productivity of the Gulf and local nutrient behaviour may affect both natural inshore fisheries and aquaculture facilities. Two topical examples show the need for more intensive study of small bays and estuaries. The recent poisoning of consumers by mussels from Prince Edward Island estuaries is now thought to have been due to a natural toxin produced by a phytoplankton. Studies will have to examine the fate of this and other toxins — how they are transported, what chemical phases they are associated with, and whether they are chemically stable in seawater. Understanding the processes that lead to the production of these toxins will also require intensive studies of inshore processes. The study of nutrient and dissolved oxygen will be one aspect of an integrated study that will require physical oceanographic expertise to determine the residence times of water at these sites and biological expertise to study the phytoplankton dynamics. Similar studies will be required to evaluate how aquaculture installations interact with the natural system and whether changes that they produce in the concentrations of nutrients, dissolved oxygen, organic carbon or other chemical species will alter the productivity of either the aquaculture facility or any adjacent wild fisheries, or produce other changes in the environment. The aquaculture industry has been growing relatively rapidly in Canada in recent years. The southern side of the Gulf from the Gaspe peninsula to Cape Breton Island and the coast of Prince Edward Island are likely to be prime sites for future aquaculture developments.

The Department of Fisheries and Oceans currently advocates that activities which affect the marine environment should be approved only when there is "no net loss of habitat". Meeting the objectives of such a policy requires a scientific competence in a variety of disciplines, including chemical oceanography/geochemistry. The continuing study of the behaviour of chemicals in the Gulf will provide one part of the expertise necessary for the pursuit of these objectives to have the best chance of success.

References

BEWERS, J. M., AND P. A. YEATS. 1977. Oceanic residence times of trace metals. Nature (Lond.) 268: 595-598.

1983. Transport of metals through the coastal zone, p. 146-163. *In* J. B. Pearce [ed.] Reviews of water quality and transport of materials in coastal and estuarine waters. Int. Coun. Explor. Sea Cooperative Res. Rep. 118, Copenhagen.

BUGDEN, G. L. 1981. Salt and heat budgets for the Gulf of St. Lawrence. Can. J. Fish. Aquat. Sci. 38: 1153-1167.

EL-SABH, M. I. 1975. Transport and currents in the Gulf of St. Lawrence. Bedford Inst. Oceanogr. Rep. BI-R-75-9: 180 p.

1977. Oceanographic features, currents, and transport in Cabot Strait. J. Fish. Res. Board Can. 34: 516-528.

LORING, D. H. 1975. Mercury in the sediments of the Gulf of St. Lawrence. Can. J. Earth Sci. 12: 1219-1237.

PETRIE, B., B.TOULANY, AND C. GARRETT. 1988. The transport of water, heat and salt through the Strait of Belle Isle. Atmos. Ocean 26: 234-251.

SUNDBY, B. 1974. Distribution and transport of suspended particulate matter in the Gulf of St. Lawrence. Can. J. Earth Sci. 11: 1517-1533.

UTHE, J., C.L. CHOU, D.H. LORING, R.T.T. RANTALA, J.M. BEWERS, J. DALZIEL, P.A. YEATS, AND R.L. CHARRON. 1986. Effect of waste treatment at a lead smelter on cadmium levels in American lobster (*Homarus americanus*), sediments and seawater in the adjacent coastal zone. Mar. Pollut. Bull. 17: 118-123.

WHITEHOUSE, B. G., G. PETRICK, AND M. EHRHARDT. 1986. Crossflow filtration of colloids from Baltic Sea water. Water Res. 20: 1599-1601.

YEATS, P. A., AND J. M. BEWERS. 1983. Potential anthropogenic influences on trace metal distributions in the North Atlantic. Can. J. Fish. Aquat. Sci. 40(Suppl. 2): 124-131.

YEATS, P. A., J. M. BEWERS, AND A. WALTON. 1978. Sensitivity of coastal waters to anthropogenic trace metal emissions. Mar. Pollut. Bull. 9: 264-268.

APPENDIX

BIO Chemical Oceanographic Data

This appendix contains information on the major oceanographic cruises (1970–84) during which chemical oceanographic sampling was conducted by members of the Chemical Oceanography Division of the Atlantic Oceanographic Laboratory (now the Marine Chemistry Division of Physical and Chemical Sciences Branch) at the Bedford Institute of Oceanography. This compilation has significant omissions: e.g. cruises not listed here formed much of the data base for the sediment geochemistry discussed in Chapter VII (see Fig. VII.1 for the sampling locations); the river monitoring program discussed in Chapters IV and V is not included; no information is given on the sediment cores discussed in Chapter VIII; additional samples were collected for many programs as circumstances allowed.

The description of each cruise consists of a figure or figures showing station positions for the cruise and a table listing the analyses performed on samples collected. For the sake of clarity, some stations have been omitted from the figures when large numbers of stations were occupied within small areas. The tables list all the parameters measured on a cruise, without regard to the total number of analyses or whether data is available at all stations and/or depths. Further information on data coverage is available from

Marine Chemistry Division
Department of Fisheries and Oceans (PCS)
P.O. Box 1006
Dartmouth, N.S. B2Y 4A2
CANADA

Requests for the analytical data may be sent to the same address.

CRUISE 70-026

Dates: July 15-July 22, 1970

Number of stations: 10

Parameters Measured: Dissolved/dispersed petroleum

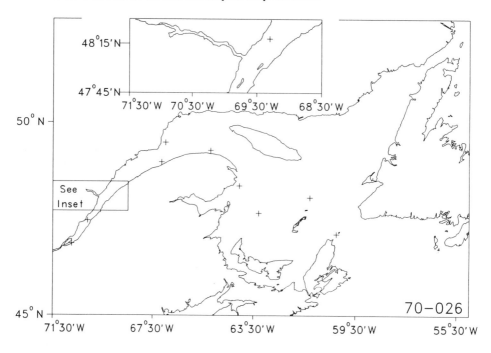

CRUISE 71-027

Dates: July 13-August 7, 1971

Number of stations: 107

Number of samples: 653

Parameters Measured: Salinity
                                Temperature
                                Dissolved oxygen

Nutrients: Silicate
                                Phosphate
                                Nitrate + Nitrite

Major ions: Li
                                Ca
                                Mg
                                Sr

Carbonate system: Alkalinity
                                Carbonate
                                pH
                                pK_a

Oil residues: Dissolved/dispersed petroleum
                                Floating particulate residues

Organic matter:	Particulate organic carbon
	Particulate organic hydrogen
	Particulate organic nitrogen

Surface sediments:	Organic carbon
	Organic nitrogen
	Lignin
	Inorganic carbon

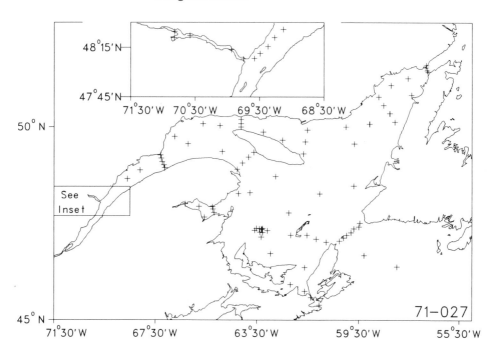

CRUISE 72-017

Dates: May 30-June 13, 1972

Number of stations: 102

Number of samples: 684

Parameters Measured: Salinity
Temperature
Dissolved oxygen

Nutrients: Silicate
Phosphate
Nitrate + Nitrite

Major ions: Li
Ca
Mg
Sr
F

Carbonate system: Alkalinity
Carbonate
pH
pK_a

| Oil residues: | Dissolved/dispersed petroleum |
| | Floating particulate residues |

Oil residues: Dissolved/dispersed petroleum
Floating particulate residues

Organic matter: Particulate organic carbon
Particulate organic hydrogen
Particulate organic nitrogen

Trace Metals: Fe (total), Fe (dissolved)
Co (total), Co (dissolved)
Ni (total), Ni (dissolved)
Cu (total), Cu (dissolved)
Zn (total), Zn (dissolved)
Cd (total), Cd (dissolved)
Pb (total), Pb (dissolved)

Surface sediments: Organic carbon
Organic nitrogen
Lignin

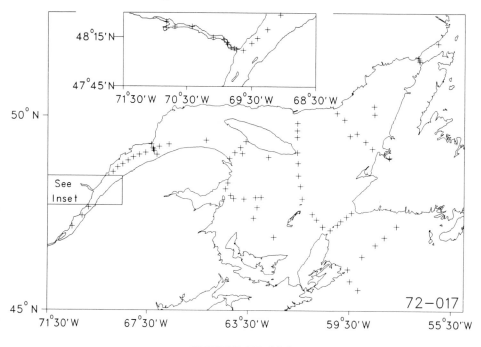

CRUISE 73-004

Dates: February 14-March 8, 1973

Number of stations: 44

Number of samples: 238

Parameters Measured: Salinity
Temperature
Dissolved oxygen

Nutrients: Silicate
Phosphate
Nitrate + Nitrite

Oil residues: Dissolved/dispersed petroleum
Floating particulate residues

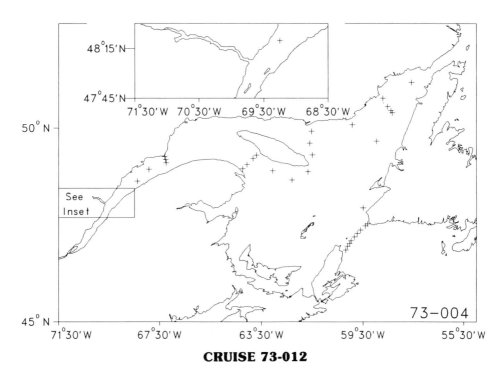

CRUISE 73-012

Dates:	April 25–May 9, 1973
Number of stations:	85
Number of samples:	398
Parameters Measured:	Salinity Temperature Dissolved oxygen
Nutrients:	Silicate Phosphate Nitrate + Nitrite
Carbonate system:	Alkalinity Carbonate pH pK_a $\delta^{13}C_{PDB}$ (total dissolved CO_2)
Oil residues:	Dissolved/dispersed petroleum Floating particulate residues
Organic matter:	Particulate organic carbon Particulate organic hydrogen Particulate organic nitrogen
Trace Metals:	Fe (total) Mn (total) Co (total) Ni (total) Cu (total) Zn (total) Cd (total) Pb (total)

Other: Suspended particulate matter
                           $\delta^{18}O_{SMOW}$ (water)

Surface sediments: Organic carbon
                           Organic nitrogen
                           Lignin
                           $\delta^{13}C_{PDB}$ (organic carbon)

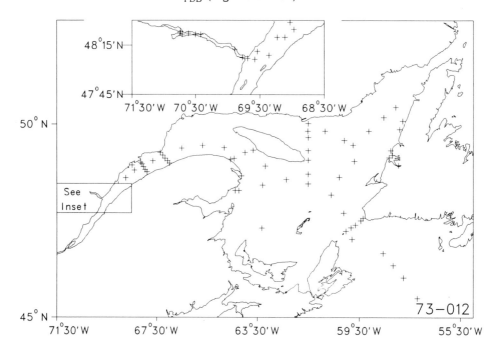

CRUISE 73-020

Dates: July 8, 1973

Number of stations: 7

Number of samples: 24

Parameters Measured:

    Oil residues: Dissolved/dispersed petroleum
                           Floating particulate residues

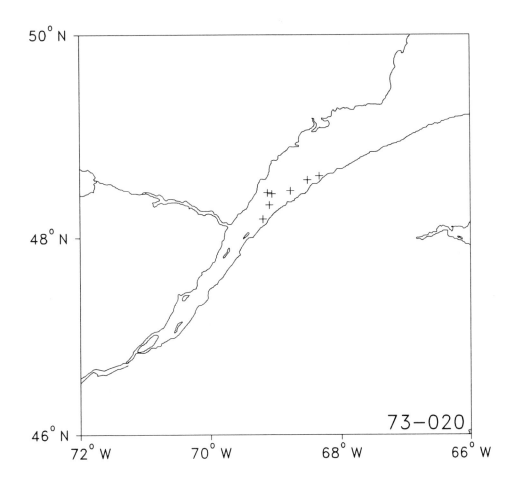

48° N

50° N

46° N

72° W   70° W   68° W   66° W

73-020

CRUISE 74-006

Dates:	April 29-May 13, 1974
Number of stations:	89
Number of samples:	696
Parameters Measured:	Salinity Temperature Dissolved oxygen
Nutrients:	Silicate Phosphate Nitrate + Nitrite
Major ions:	F
Carbonate system:	Alkalinity pH

Trace Metals: Fe (total), Fe (dissolved)
                          Mn (total), Mn (dissolved)
                          Co (total), Co (dissolved)
                          Ni (total), Ni (dissolved)
                          Cu (total), Cu (dissolved)
                          Zn (total)
                          Cd (total)
                          Hg (dissolved), Hg (particulate)

Other: Suspended particulate matter
                          $\delta^{18}O_{SMOW}$ (water)

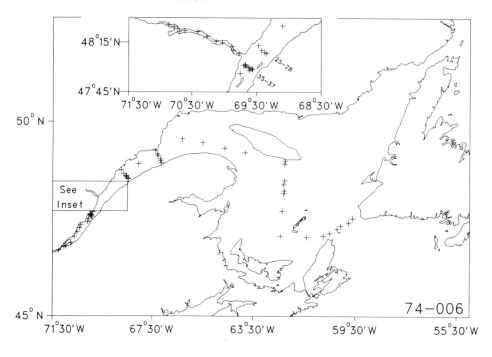

CRUISE 74-028

Dates: July 30–August 14, 1974

Number of stations: 91

Number of samples: 928

Parameters Measured: Salinity
                             Temperature
                             Dissolved oxygen

Nutrients: Silicate
                             Phosphate
                             Nitrate + Nitrite

Carbonate system: Alkalinity
                             Carbonate
                             pH
                             pK_a
                             $\delta^{13}C_{PDB}$ (total dissolved CO_2)

| Oil residues: | Dissolved/dispersed petroleum |
| | Floating particulate residues |

Organic matter:	Particulate organic carbon
	Particulate organic hydrogen
	Particulate organic nitrogen
	$\delta^{13}C_{PDB}$ (particulate organic carbon)

| Other: | Suspended particulate matter |

Surface sediments:	Organic carbon
	Organic nitrogen
	Lignin
	$\delta^{13}C_{PDB}$ (organic carbon)

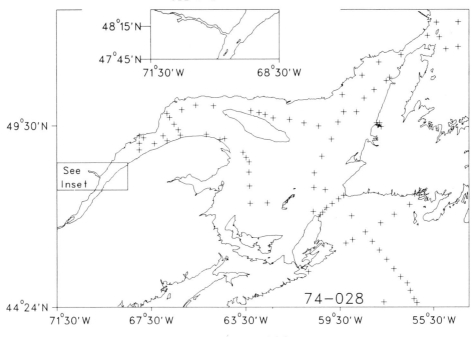

CRUISE 74-032

Dates:	September 18–September 29, 1974
Number of stations:	104
Number of samples:	803

Parameters Measured:	Salinity
	Temperature
	Dissolved oxygen

| Oil residues: | Dissolved/dispersed petroleum |
| | Floating particulate residues |

Organic matter:	Particulate organic carbon
	Particulate organic hydrogen
	Particulate organic nitrogen

Surface sediments:	Organic carbon
	Organic nitrogen
	Lignin
	$\delta^{13}C_{PDB}$ (organic carbon)

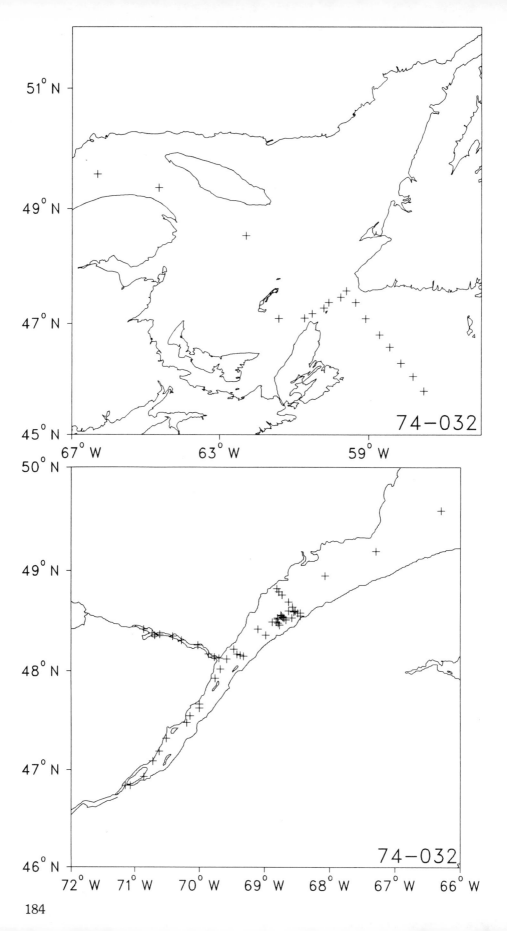

74-032

74-032

CRUISE 75-015

Dates: May 29-June 7, 1975

Number of stations: 60

Number of samples: 465

Parameters Measured: Salinity
Temperature
Dissolved oxygen

Nutrients: Silicate
Phosphate
Nitrate + Nitrite

Carbonate system: Alkalinity
Carbonate
pH
pK_a
$\delta^{13}C_{PDB}$ (total dissolved CO_2)

Oil residues: Dissolved/dispersed petroleum
Floating particulate residues

Organic matter: Dissolved organic carbon
Particulate organic carbon
Particulate organic hydrogen
Particulate organic nitrogen

Other: Suspended particulate matter
Halogenated hydrocarbons
$\delta^{18}O_{SMOW}$ (water)

Surface sediments: Organic carbon
Organic nitrogen
Lignin
$\delta^{13}C_{PDB}$ (organic carbon)

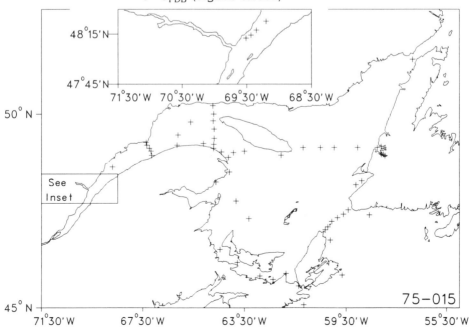

CRUISE 75-031

Dates: November 17-28, 1975

Number of stations: 6

Number of samples: 50

Parameters Measured: Salinity
                       Temperature
                       Dissolved oxygen

Organic matter: Dissolved organic carbon
                       Particulate organic carbon
                       Particulate organic hydrogen
                       Particulate organic nitrogen

Other: Suspended particulate matter

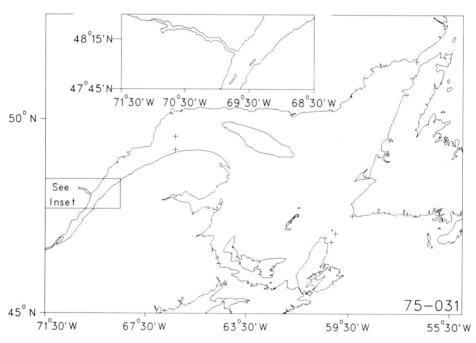

CRUISE 76-006

Dates: April 23-April 28, 1976

Number of stations: 70

Number of samples: 340

Parameters Measured: Salinity
                       Temperature
                       Dissolved oxygen

Oil residues: Dissolved/dispersed petroleum
                       Floating particulate residues

Organic matter:	Particulate organic carbon
	Particulate organic hydrogen
	Particulate organic nitrogen

Carbonate system: $\delta^{13}C_{PDB}$ (total dissolved CO_2)

Trace Metals:	Fe (dissolved), Fe (particulate)
	Mn (dissolved), Mn (particulate)
	Co (dissolved)
	Ni (dissolved)
	Cu (dissolved), Cu (particulate)
	Zn (dissolved), Zn (particulate)
	Cd (dissolved), Cd (particulate)
	Al (dissolved), Al (particulate)
	Ca (particulate)

Other:	Suspended particulate matter
	$\delta^{18}O_{SMOW}$ (water)

Surface sediments:	Organic carbon
	Organic nitrogen
	Lignin
	$\delta^{13}C_{PDB}$ (organic carbon)

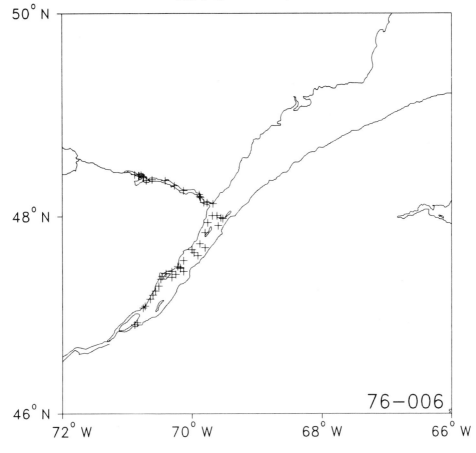

CRUISE 76-021

Dates:	June 25-June 30, 1976
Number of stations:	59

Number of samples: 640

Parameters Measured: Salinity
Temperature
Dissolved oxygen

Nutrients: Silicate
Phosphate
Nitrate + Nitrite

Carbonate system: $\delta^{13}C_{PDB}$ (total dissolved CO_2)

Oil residues: Dissolved/dispersed petroleum
Floating particulate residues

Trace Metals: Fe (total), Fe (dissolved)   Cu (total)
Mn (total), Mn (dissolved)   Zn (total)
Co (total)   Cd (total)
Ni (total)

Other: Suspended particulate matter
$\delta^{18}O_{SMOW}$ (water)

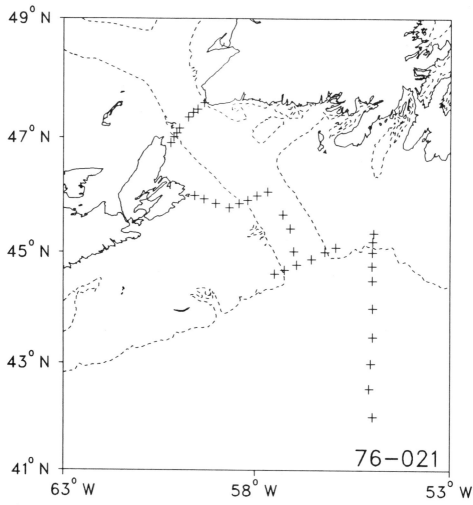

The dashed line on the figure is the 200 m depth contour.

CRUISE 79-024

Dates:	August 25–October 10, 1979
Number of stations:	58
Number of samples:	451
Parameters Measured:	Salinity
	Temperature
Carbonate system:	$\delta^{13}C_{PDB}$ (total dissolved CO_2)
Oil residues:	Dissolved/dispersed petroleum
	Floating particulate residues
	Surface microlayer
Organic matter:	$\delta^{13}C_{PDB}$ (particulate organic carbon)
	$\delta^{13}C_{PDB}$ (plankton)
Trace Metals:	Fe (dissolved), Fe (particulate)
	Mn (dissolved), Mn (particulate)
	Co (dissolved)
	Ni (dissolved)
	Cu (dissolved), Cu (particulate)
	Zn (dissolved), Zn (particulate)
	Cd (dissolved), Cd (particulate)
	Al (dissolved), Al (particulate)
	Ca (particulate)
	V (dissolved)
	Cr (dissolved)
	Ba (dissolved)
Other:	Suspended particulate matter
Surface sediments:	Organic carbon
	$\delta^{13}C_{PDB}$ (organic carbon)

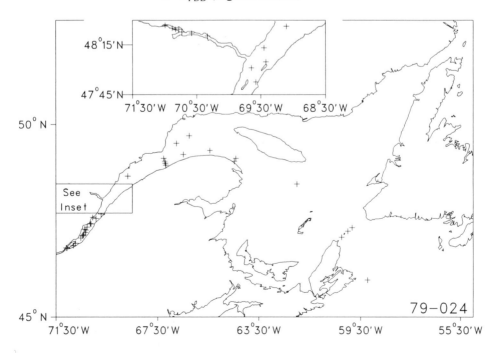

CRUISE 84-001

Dates: January 25-February 2, 1984

Number of stations: 47

Number of samples: 350

Parameters Measured: Salinity
Temperature
Dissolved oxygen

      Nutrients: Silicate
Phosphate
Nitrate + Nitrite

      Carbonate system: Alkalinity
Carbonate
pK_a

      Organic matter: Particulate organic carbon
Particulate organic nitrogen
Total organic carbon

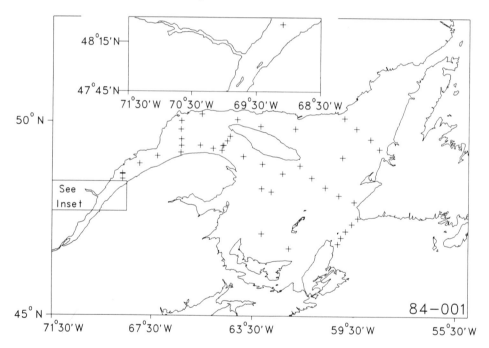